度假村与酒店
Resort Hotel

韩国建筑世界出版社 编
王单单 李硕 闫文诗 李群 周杰姝 曲艺 王洋 译

大连理工大学出版社

目录 CONTENTS

度假村 RESORT

酒店 HOTEL

nhow酒店 nhow Hotel
德国 Germany

圣托里尼·格雷斯酒店
Santorini Grace Phase
希腊 Greece

未来酒店 Future Hotel
德国 Germany

KUUM温泉住宅酒店
The Kuum Hotel Spa &
Residences
土耳其 Turkey

安宁河水上休闲度假中心
Anning River Aquatic
Recreation Resort
中国 China

江苏金诚酒店与度假村
Jiangsu Gold Sense Hotel & Resort
中国 China

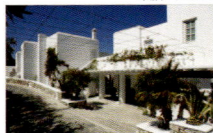
观景楼酒店
Belvedere Hotel
希腊 Greece

圣彼得堡城市度假村
Urban Resort Saint
Petersburg
俄罗斯 Russia

杜塞尔多夫凯悦酒店
Hyatt Regency Dusseldorf
德国 Germany

中国威海锦湖韩亚高尔夫俱乐部
China Weihai Point Hotel & Golf
Resort
中国 China

Art Del Teatre酒店
Hotel Arc Del Teatre
西班牙 Spain

瓦地度假村 Wadi Resort
约旦 Jordan

伦敦W酒店 W Hotel London
英国 United Kingdom

Orhidelia大型健康休闲中心
Wellness Orhidelia
斯洛文尼亚 Slovenia

Isla Moda 度假村
Isla Moda
阿拉伯联合酋长国
United Arab Emirates

希尔顿芭提雅酒店
Hilton Pattaya
泰国 Thailand

Gele海滩与度假村
Gele Beach and Resort
尼日利亚 Nigeria

宿务岛皇居水上乐园度假村
Imperial Palace Waterpark Resort & Spa,
Cebu
菲律宾 Philippines

Saffire度假村 Saffire Resort
澳大利亚 Australia

南港度假屋
South Harbor Resort
日本 Japan

圆形别墅 Villa Ronde
日本 Japan

Bayside Marine酒店
Bayside Marina Hotel
日本 Japan

Utoco深海治疗中心与酒店
Utoco Deep Sea Therapy
Center & Hotel
日本 Japan

Amangiri度假村 Amangiri
美国 America

艾姆斯酒店 Ames Hotel
美国 America

Daemyung Sonofelice 度假
俱乐部
Daemyung Sonofelice Resort Club House
韩国 Korea

韩华金海湾高尔夫度假村
Hanhwa Golden Bay Golf & Resort
韩国 Korea

仁川喜来登酒店
Sheraton Incheon Hotel
韩国 Korea

济州岛凤凰度假村 Vella Terrace
公寓
JEJU Phoenix Island Vella Terrace
韩国 Korea

玛雅里维埃拉东方酒店
Mandarin Oriental R verea
Maya
墨西哥 Mexico

墨西哥度假村 Mexico Resort
墨西哥 Mexico

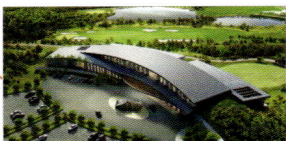
MOW俱乐部
CLUB MOW Club House
韩国 Korea

Rock It Suda Rock It Suda
韩国 Korea

海岬酒店 Headlands Hotel
澳大利亚 Australia

酒店　HOTEL

度假村 RESORT

宿务岛皇居水上乐园度假村　Imperial Palace Waterpark Resort & Spa, Cebu

度假村 RESORT

Orhidelia 大型健康休闲中心
Wellness Orhidelia

地点　Podcetrtek, Slovenia
设计　ENOTA
功能　Health facility
总楼面面积　9,990m^2
参与设计人员　Maruša Zupancic, Nuša Završnik, Zana Starovic, Anna Kravcova, Polona Ruparcic, Marko Volf, Sabina Sakelšek, Esta Matkovic, Darja Zubac, Dean Jukic, Tjaša Marinšek, Nebojša Vertovšek, Nom biro(mechanical services), Forte inženiring(electrical planning), Darrtech(pool technology)
甲方　Terme Olimia
景观设计　Bruto
结构工程师　Elea iC
摄影师　Miran Kambic

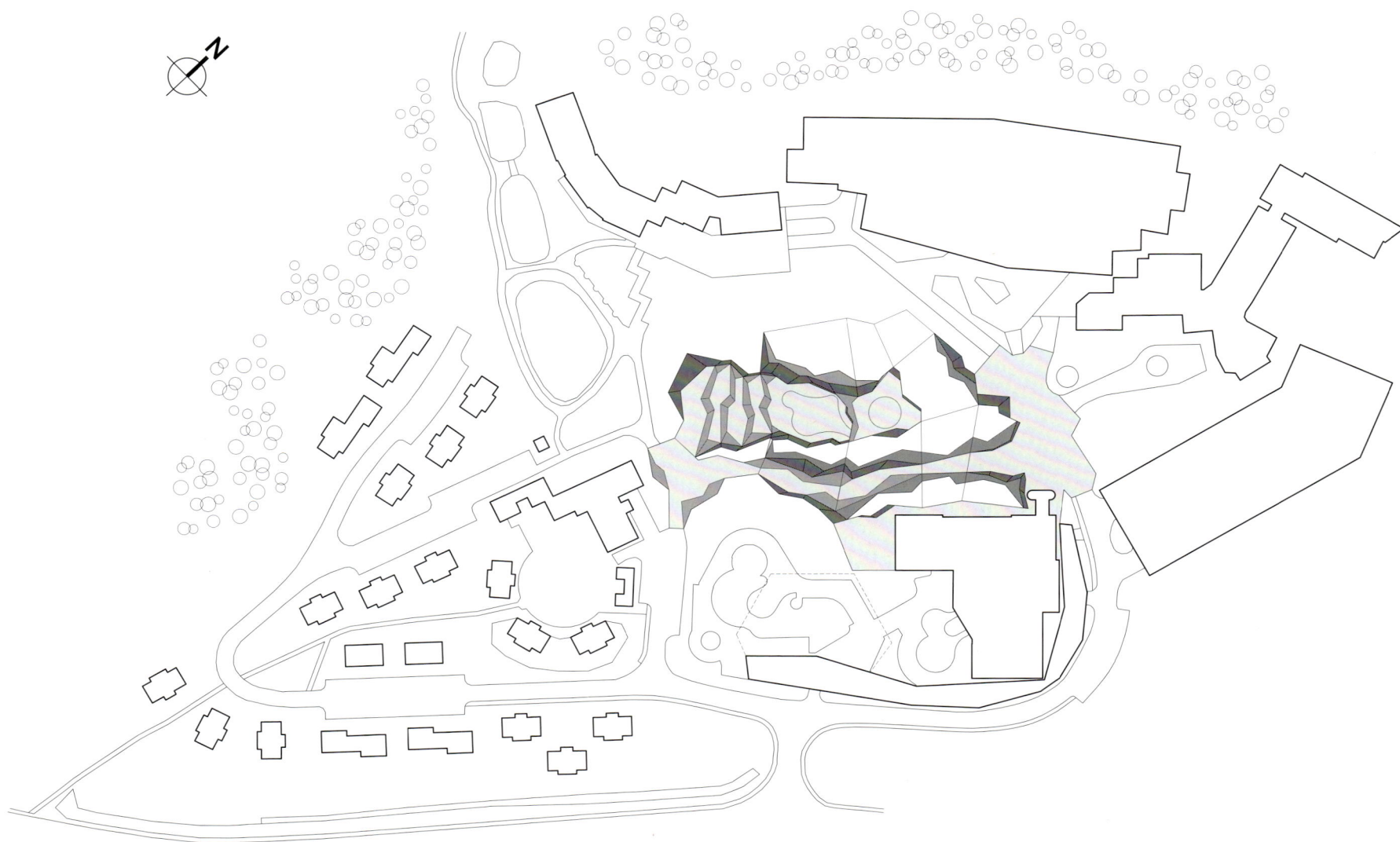

总平面图 Site plan

设计师设计该项建筑的主要目的是使它尽可能地融入到周围环境当中。由于健康休闲中心的内容广泛多样,有些部分就需要克服内部空间的巨大跨度和高度的限制,如果采用传统的方式,在中心绿色区域安置建筑,就会把剩下的开放空间填满,大大降低空间的特征。

新的健康休闲中心的设计方式是连续的,它更像一个景观,而非建筑。曲折的立面好像一面支承墙,将景观表面分为不同等级。中间的步行道现在已经延伸到了屋顶,给游客带来全新的体验。在两端,建筑内、外道路的连接处形成两个较小的广场,用来控制车速,给行人带来便利。

这座新建筑并没有彰显它的独特性,而是与现有的独立建筑和周围的环境组成一个整体。

The main goal while designing the building was to diminish as much as possible its presence in the surroundings. Since the demanded program of wellness center is very extensive and in parts it demands overcoming great spans and big heights of inner spaces, putting up classically conceived building on central green plot will fill up last remaining open area in thermal complex and largely degrades its spatial quality.

New wellness center is consecutively designed rather like a landscape arrangement then a building. Folded elevations appear like supporting walls dividing different levels of landscape surfaces. Central walking path is now stretched over the roof and enables visitor completely new, different experience of the site. On both ends, where strolling path connects with passing inner roads, it forms two smaller public squares to control the speed of vehicles and ultimately gives advantage to pedestrians over the traffic.

路径 1 Trail 1

开放空间 Open space

Rather than searching for its own expression and claiming its space new object connects existing single buildings and other spatial elements on the whole.

路径2 Trail II

示意图 Diagram

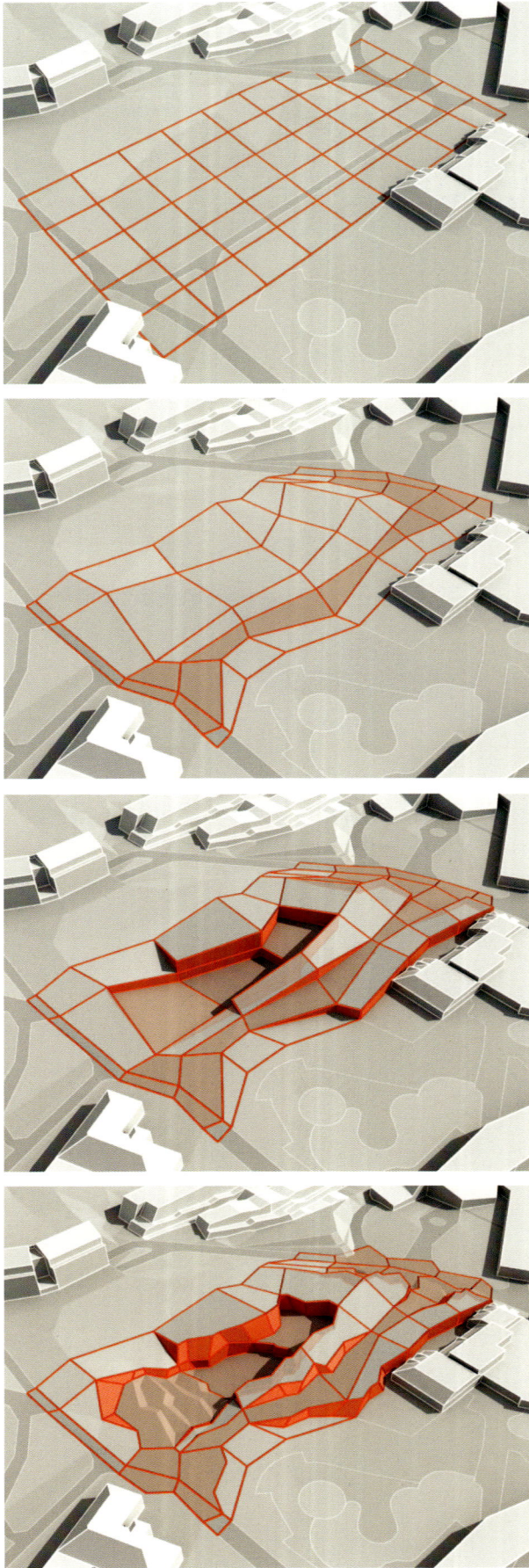

开放空间（夜景） Open space(night)

立面详图 Detailed elevation

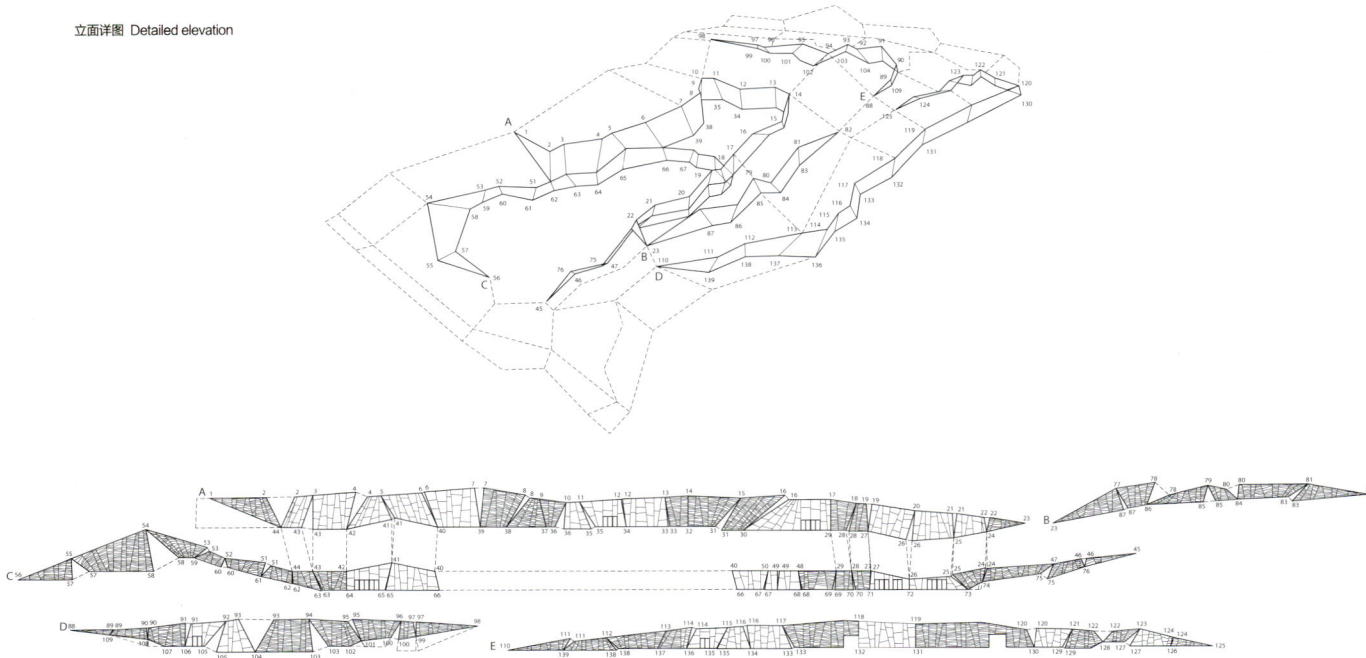

外观（夜景） Exterior(night)

柱子设计图 Column scheme

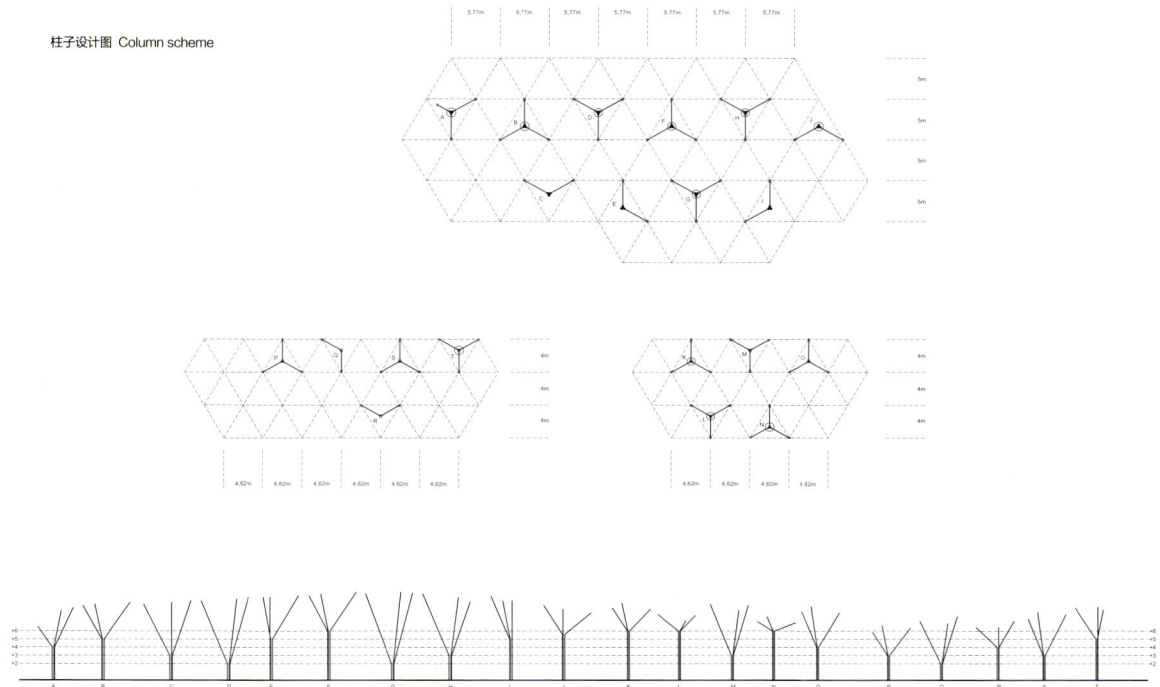

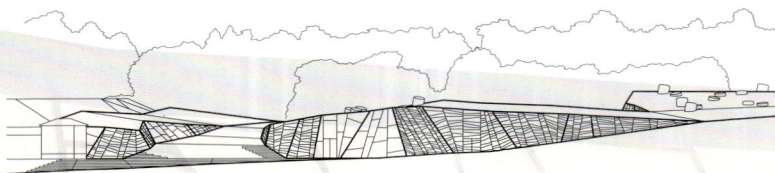

北立面 North elevation

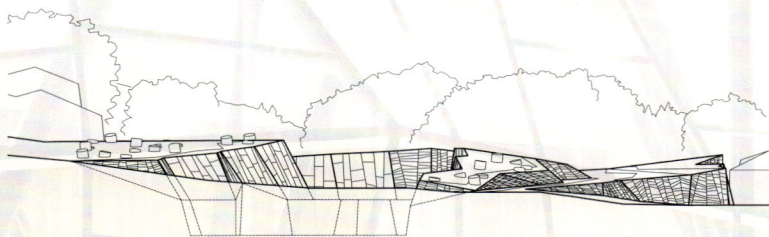

南立面 South elevation

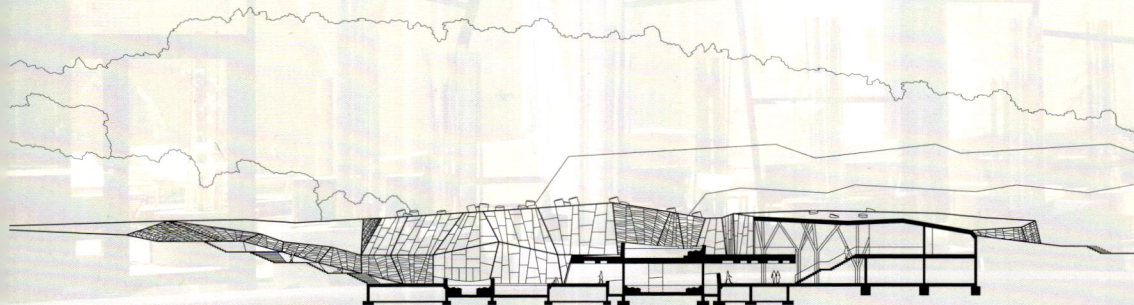

纵剖面 Longitudinal section

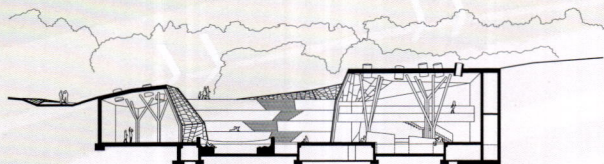

横剖面 Cross section

室内 4 Interior IV

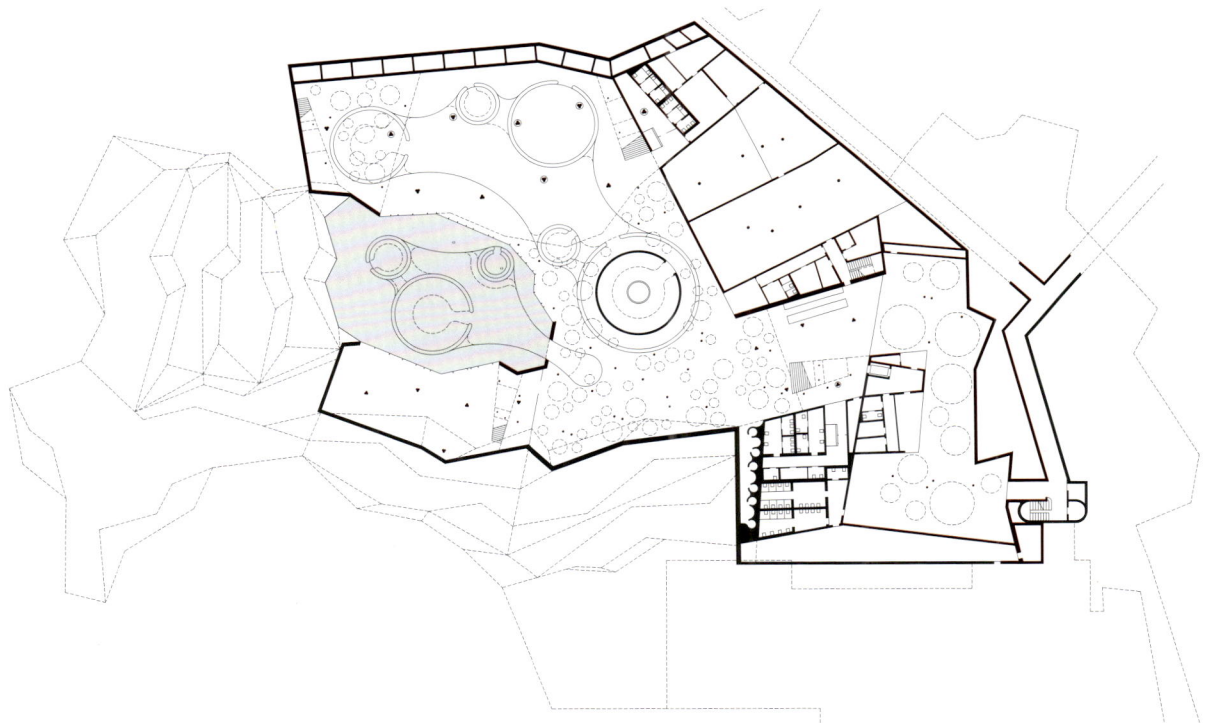

一层平面图 First floor plan

室内 5 Interior V

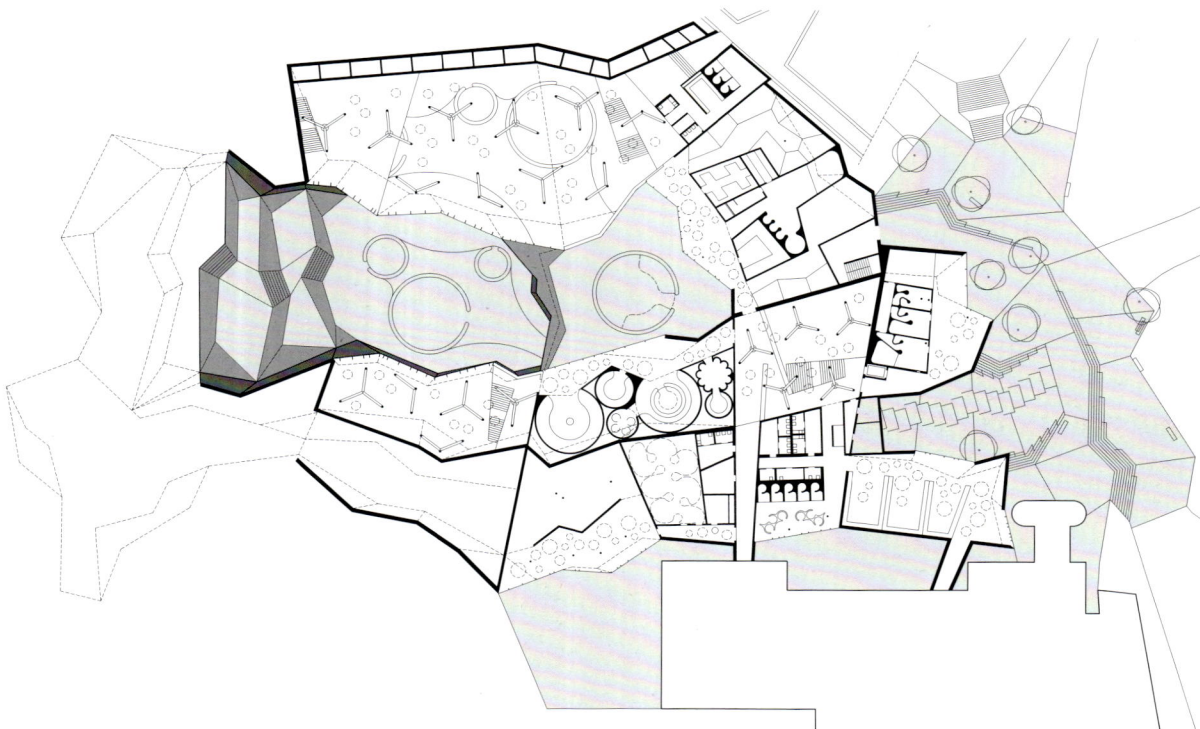

二层平面图 Second floor plan

室内 6 Interior VI

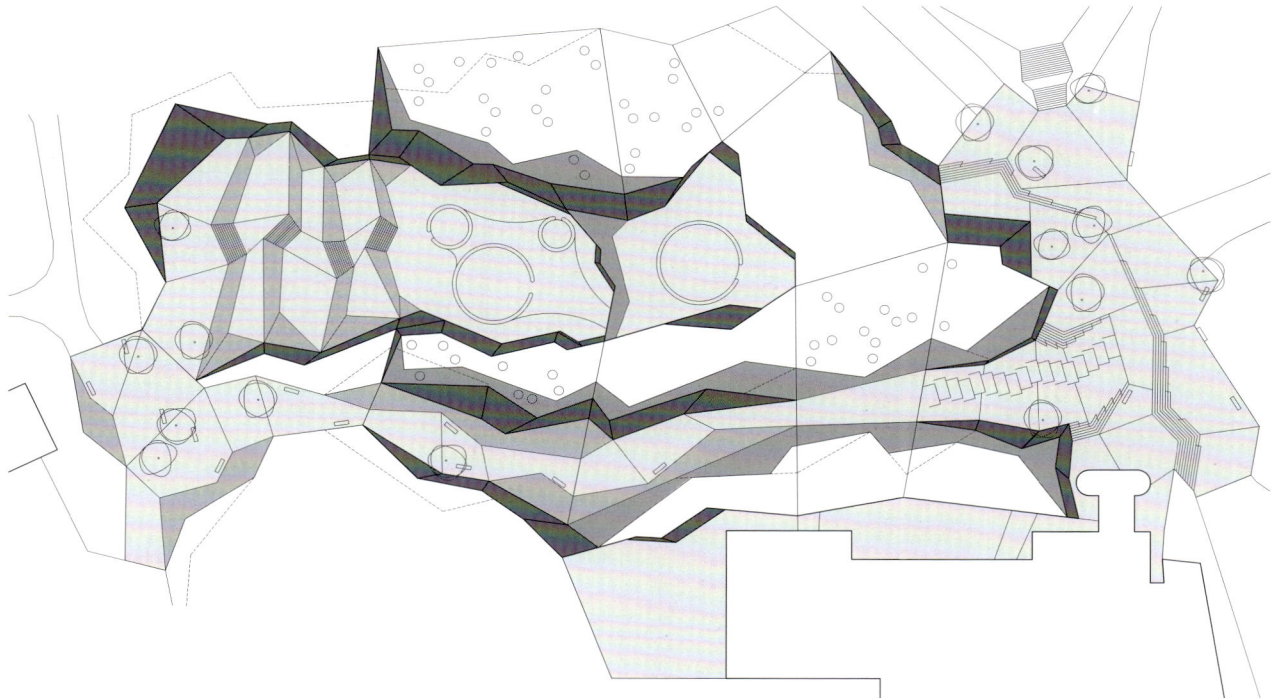

屋顶平面图 Roof floor plan

宿务岛皇居水上乐园度假村
Imperial Palace Waterpark Resort & Spa, Cebu

设计 BAKU PLAN CO.,LTD. / Seichiro Nagasawa + DESIGN US / Seo Jong-ho, Kang Min-Gyeong

地点 Mactan Island, Cebu, Philippines

建筑面积 68,695m²

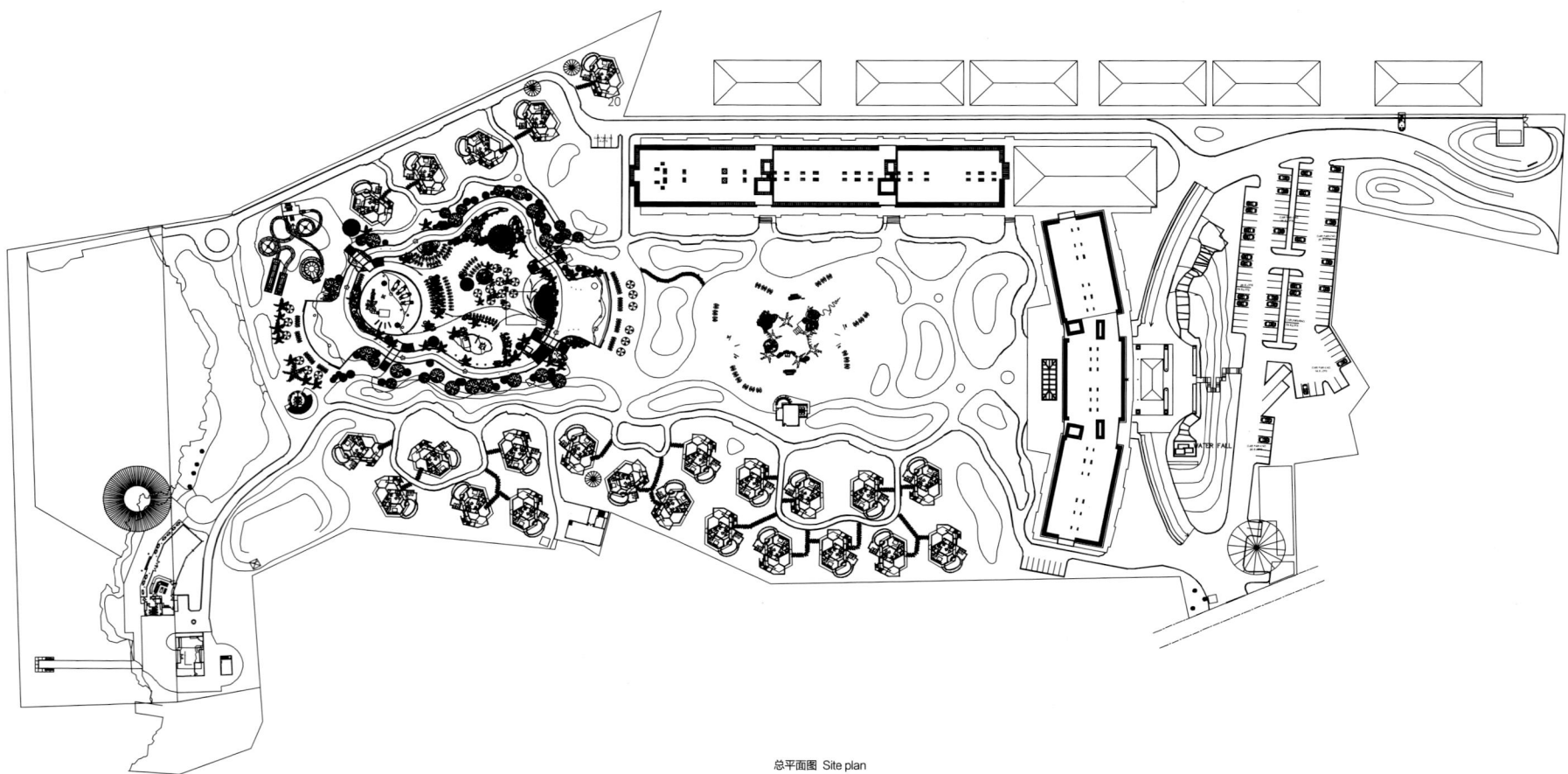

总平面图 Site plan

大堂 Lobby

卧室（泳池别墅） Bedroom(pool villa)

在去旅行的路上，我们满怀期待地进行着准备，并对一个新的旅行地点充满了好奇。皇居水上乐园度假村的设计者提出了"旅行"这个理念，旨在创建一个清新的、高品质的内部空间，这种特色在主入口、建筑入口、门厅和登记处随处可见。这种娱乐效果和带给人们的好奇感来源于各种空间的组合，它们展示了菲律宾的异国情调、西方建筑的风格和当地饰面材料的特色。

大堂是酒店给人们的第一印象，通过展现大型罗马式拱门和墙壁与圆柱上的柔和弧线，使人们感受到西班牙式的氛围。这些元素与宿务岛的海景相得益彰，增强了它的异国情调，这一点也在具有当地度假风格的温泉胜地中得以体现。温泉入口的特点是脚下池塘的一块块垫脚石，呈现出来的自然感觉在城市中很难看到。

位于水上乐园附近的泳池别墅离这个度假村较远，目的是为了保持它的独立性。这个别墅四周围绕着一个美丽的花园，游客在这里可以免受外界的打扰，感受宁静安逸与轻松自在。泳池别墅分为两层：一层是一个游泳池，二层是一个八人按摩浴缸，两层可以独立使用。

We experience anticipation while preparing for a travel and curiosity about a new place on the way to the destination. Setting "travel" as the concept for Imperial Palace Waterpark Resort, the designer intended to create an interior space with freshness and quality, which can be found along the main entrance, building entrance, lobby and check-in area. The amusement and curiosity is stimulated by the various spatial compositions and presentations featuring the exotic feel of Philippines, Western architectural style and local finish materials.

The lobby, which decides the first impression of a hotel, shows the ambience of Spain through the large Romanesque arch and the gentle curve applied to the walls and columns. These elements harmonize with the seascape of Cebu to enhance the exoticness, which is also seen in the spa finished by local materials that emit the atmosphere of a resort. The entrance of the spa is characterized by stepping stones of a pond to present a natural feeling that is rarely found in the city.

The pool villa in the vicinity of the water park is located far from the resort building in order to maintain the independency. Surrounded by a garden of a beautiful view, it provides the customers with tranquility and privacy so that they can relax comfortably. The two-story pool villa consists of a pool on the first floor and a eight-person whirlpool bath on the second floor, which can be used separately.

室内（泳池别墅） Interior(pool villa)

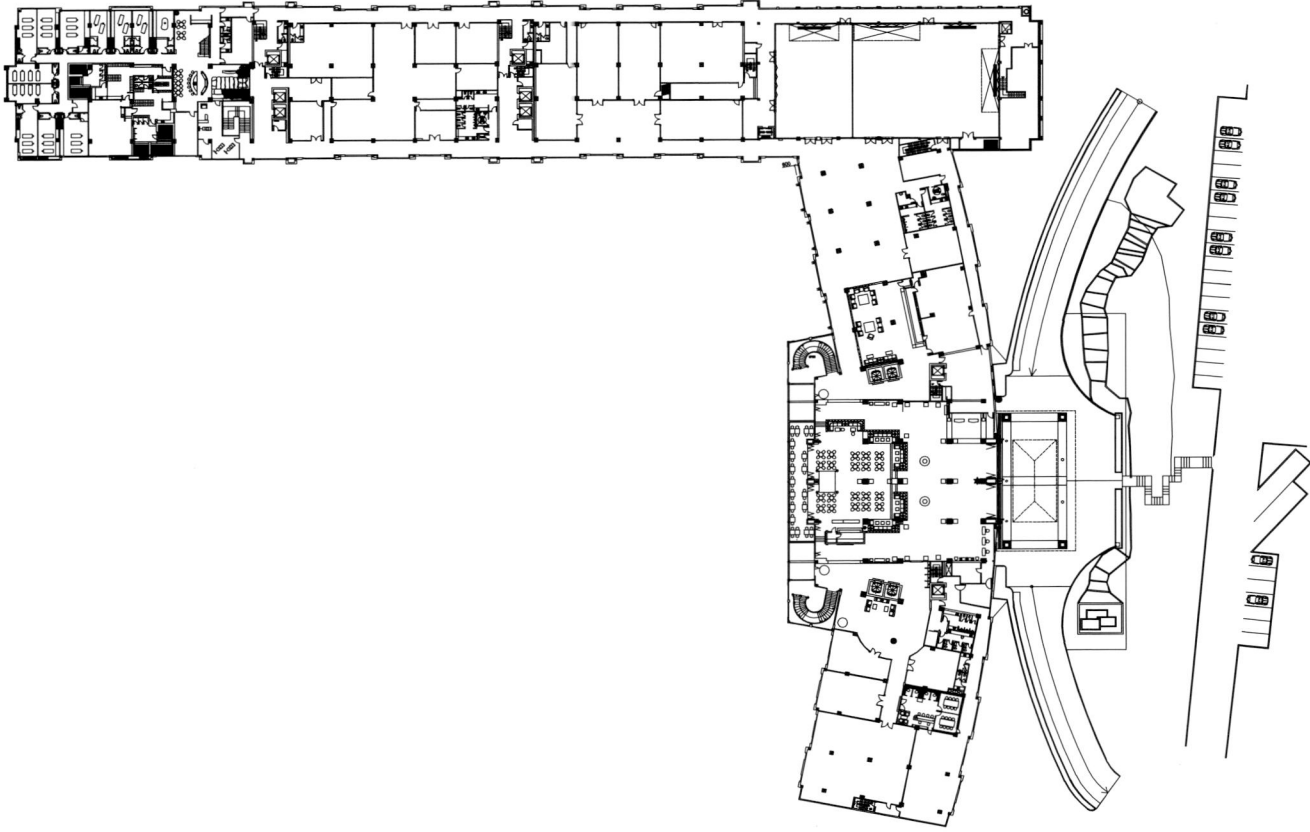

二层平面图 Second floor plan

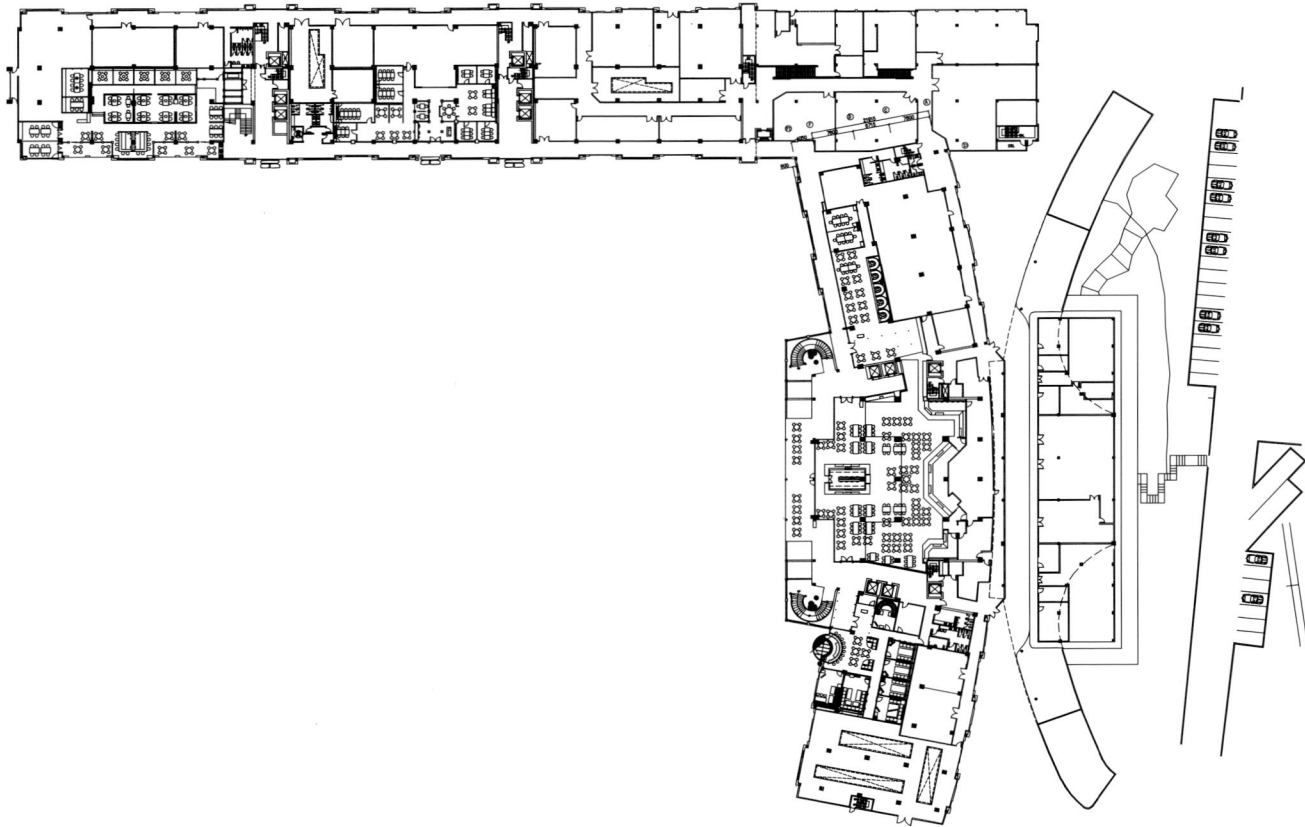

一层平面图 First floor plan

0   15   50(M)

自助餐厅与咖啡厅 1 Buffet & Cafe Ⅰ

自助餐厅与咖啡厅 2 Buffet & Cafe Ⅱ

水疗区（室内） Spa(int.)

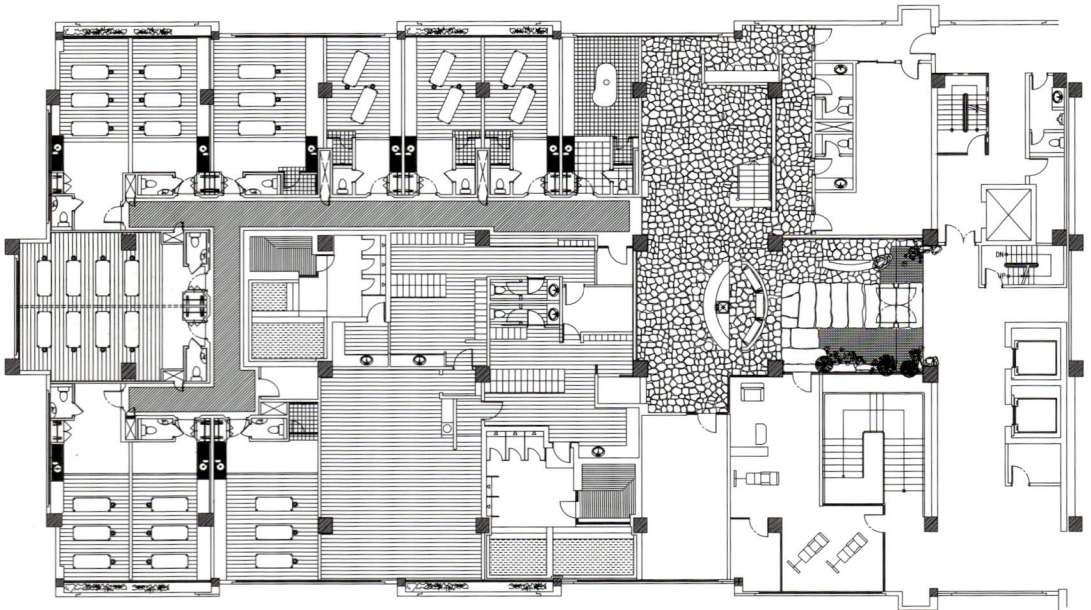

0   2   5(M)

楼层平面图 Floor plan

水疗区（室外）Spa(ext.)

房间1 Room I

房间2 Room II

泳池别墅 Pool Villa

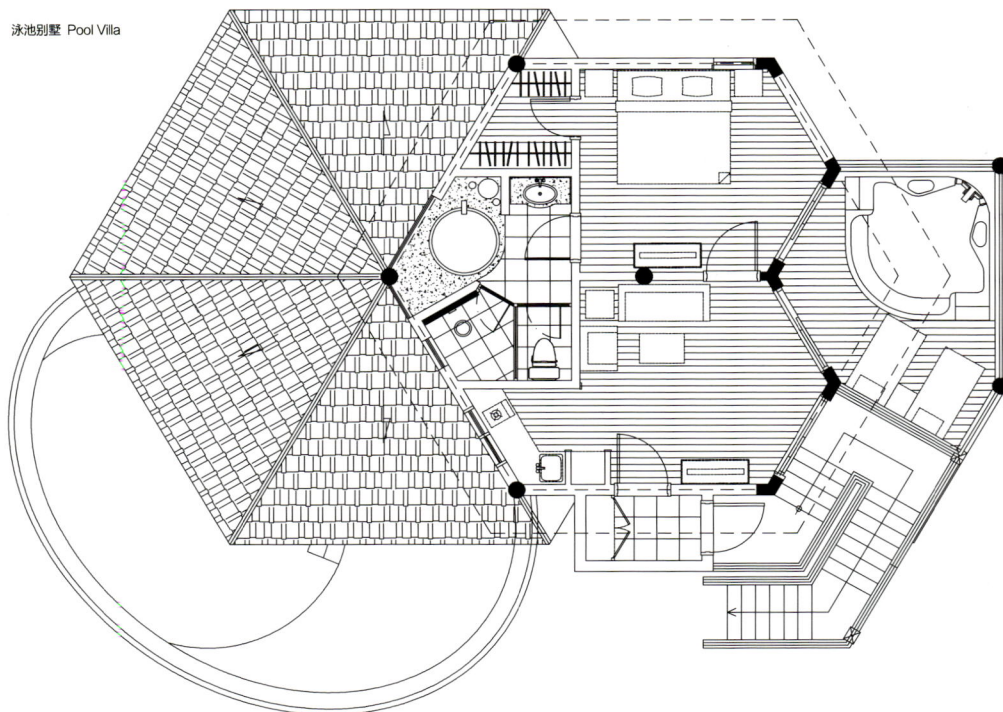

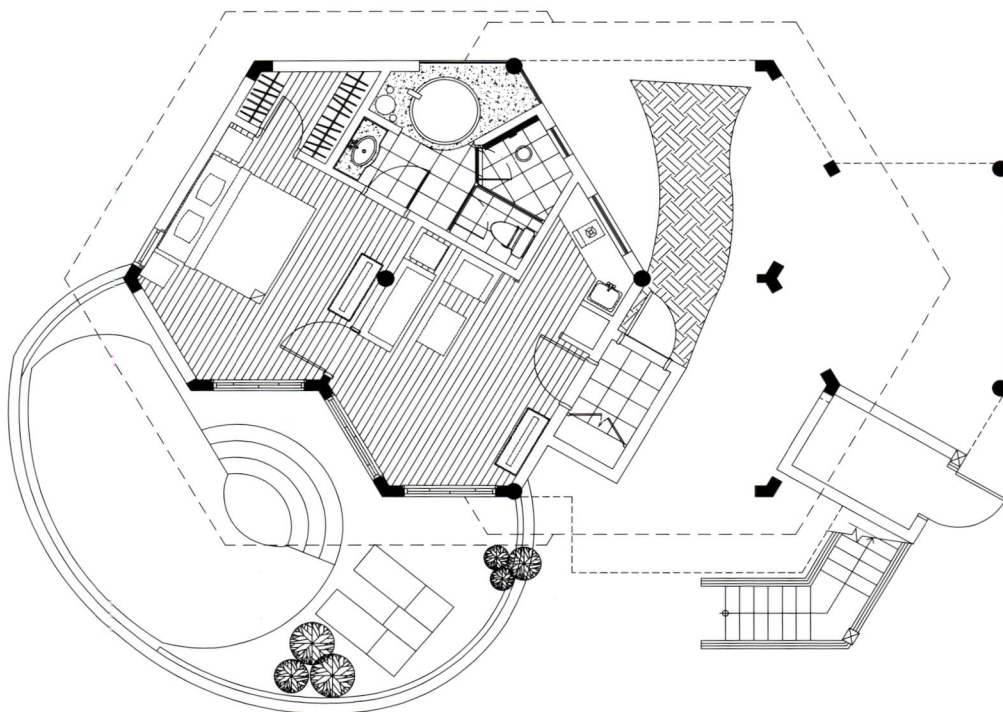

二层平面图——按摩浴缸别墅 Second floor plan (jacuzzi villa)

一层平面图——泳池别墅 First floor plan(pool villa)

0   2   5(M)

泳池别墅（夜景）　Pool villa(night)

圣托里尼·格雷斯酒店
Santorini Grace Phase

设计　mplusm
地点　Santorini, Greece
基地面积　1,033m^2
建筑面积　517m^2
甲方　Grace Hotels Group
室内设计　Sophia Vantaraki
摄影师　Serge Detalle

0   5   10(M)

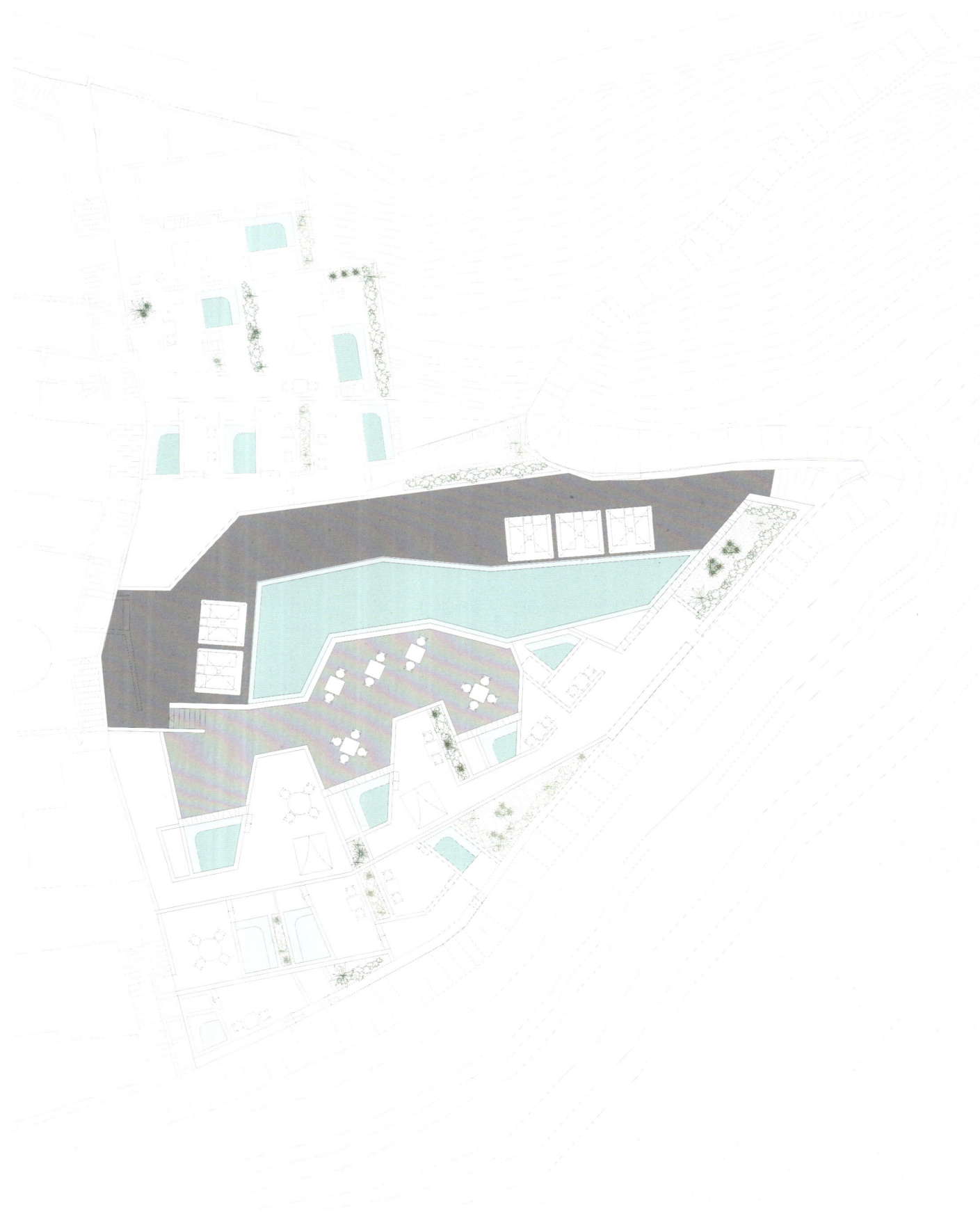

总平面图 Site plan

从泳池看 View from swimming pool

在一种特殊的背景条件下——比如圣托里尼伊莫洛维里，这里面对火山口，背对着斯卡洛斯岩石——应该如何恢复本土的建筑样式？大多伊莫洛维里的房间一直采用拱形天花板的设计，来模仿本土的风格。然而，圣托里尼格雷斯酒店并没有使用这种拱顶结构：房间的布局是按照地面上凌乱的几何形状来设计的，之所以有这样的几何形状，是因为陡峭的火山口周围形成了一系列新的礁石。这些锯齿线条汇集在斜坡的中间位置，这里是整个项目基地中宽度最大的地方，游泳池、餐厅和新酒店的酒吧都位于这里。

从外观上看，酒店的石墙产生了一个黑白相间的效果：一些高处的新客房可以看到该地区的火山岩石，这些岩石带有空隙，可以让光线穿透进入房间。这种对当地石墙的重新诠释，可以确保空气流通，保护游客的隐私。因为面向火山口是这里的一个典型的特色。在这些房间中，高处的凹形空间可以形成一个凝视海景的小型私人空间。身处房间中，你不仅看得到基克拉迪的白色装饰内壁，而且可以感受到火山岩石的正面，这是一个极其特殊的地方。厚厚的墙体是圣托里尼具有历史意义的高塔和堡垒的另一个元素，它也被引入到酒店的设计中去，这种墙体的厚度可以容纳一个类似衣柜的家具。房间里的混凝土地面保持着质朴的本色，与墙上的黑红两色石块一起与岛上独特的地形交相呼应。

最近，圣托里尼·格雷斯酒店完成了扩建，增设了一个新的餐厅、酒吧区、一个新的接待处和三间新客房。设计师设计了一种与岛上特殊地质相吻合的黑色洞穴空间，这种设计理念不仅在新的接待处得以应用，而且也体现在餐厅和酒吧里：接待处的背景墙是由黑色的火山石修建的，并带有一些引人注目的空隙，有点类似酒店一些房间的布置。餐厅和酒吧区的背景墙绘制了深色的火山石图案，其棱柱形的几何形状就像是一幅由岩石组成的风景画。这面背景墙环绕着桌子后面的长椅、展示橱窗和酒架。酒架是金属的，有几排水平放置酒瓶的圆形穿孔，这种抽象平面上的穿孔类似于黑色火山石的空隙。

How could one revive vernacular architecture in such an exceptional setting as Imerovigli, Santorini, with its view of Caldera, opposite the Skaros rock? Because generally at Imerovigli rooms to rent simulate the vernacular by cloning indefinitely vaulted ceilings. However at Santorini Grace, we have used no vaults: the layout of rooms follows in plan the broken geometries of the ground's contours, forming in effect a series of new stone ledges in the steep drop of Caldera. These jagged lines converge in mid-slope at the level which produces the greatest possible width within the site and it is at this level that the pool, the restaurant and the bar of the new hotel are situated.

On the exterior, stone walls create a rhythm of black and white for the hotel: Some of the new rooms have acquired for their elevation a face of the region's recognizable volcanic rock with gaps from which rays of light are allowed to pass. This reinterpretation of the vernacular stone walls ensures ventilation and privacy from passersby guests since rooms – a typically feature in the Caldera front – are front loaded and fully open to the view. In these rooms, the recesses of elevation form small private open spaces for contemplating the view. Being in the interior of these rooms you are not just surrounded by the usual Cycladic white interior but you also perceive the volcanic rock facade, a potent reminder of the particular place you are located. Thick walls, another element of Santorini's historic towers and fortresses, are introduced in the hotel as a wall thickness that accommodates several built in furnishings such as wardrobes and vanity. The concrete floors of the rooms retain the earthy palette of colors – together with the black and red stones in the walls they form a dialogue with the island's unique exposed geological section.

Recently, Santorini Grace completed its extension with a new restaurant and bar area, a new reception and three new rooms. The idea of a black cavern space in equivalence to the special geology of the island has been maintained in the interior not only of the new reception but also the restaurant and the bar: on the background of the reception desk a stone wall has been built from black volcanic stones with remarkable voids, resembling the ones applied to some of the rooms of the hotel. The background of the restaurant and bar area is painted in the dark color of the volcanic stones and its prismatic geometry resembles a rocky landscape. This background embraces the long seating bench behind the tables, the facade of the show kitchen window as well as the wine rack. The wine rack is metallic and equipped with rows of circular holes for the horizontal placement of the bottles, holes that on an abstract level resemble the gaps between the black volcanic stones.

从露台看 View from terrace

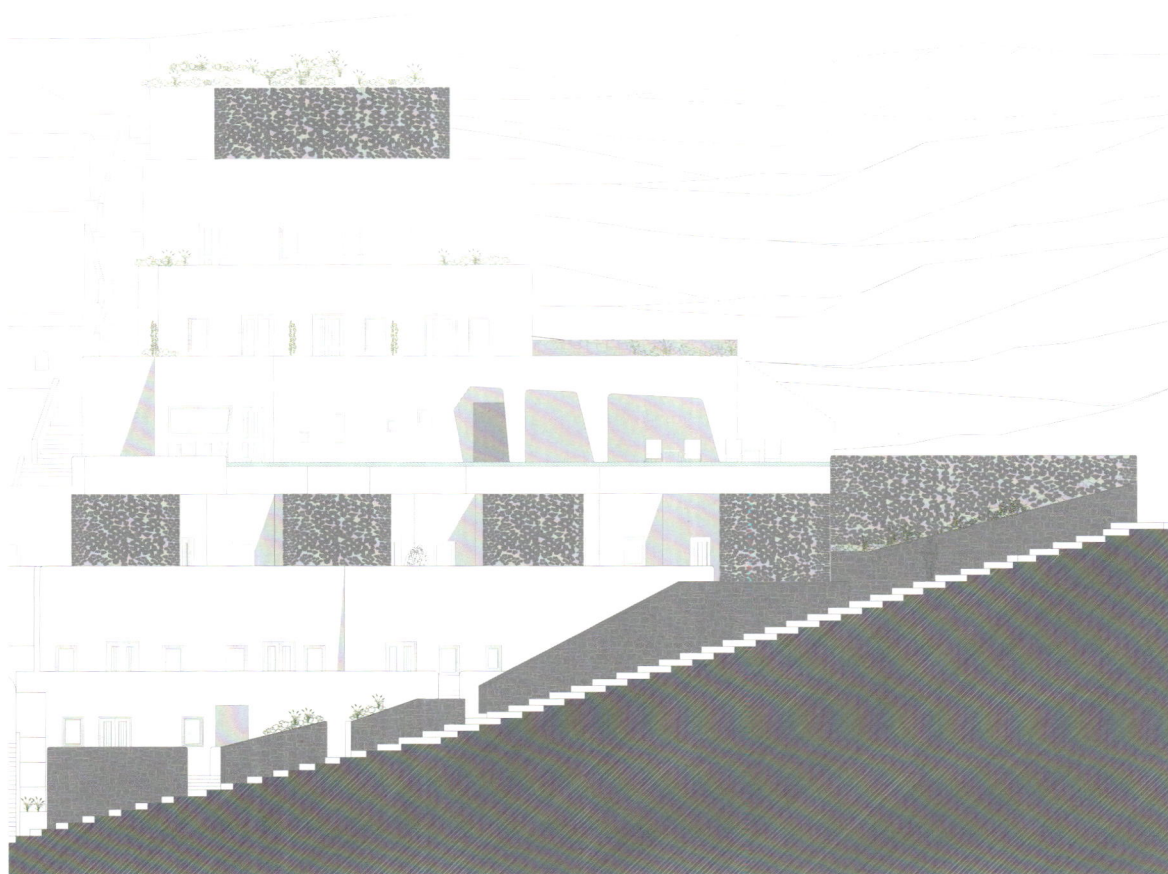

立面图 Elevation

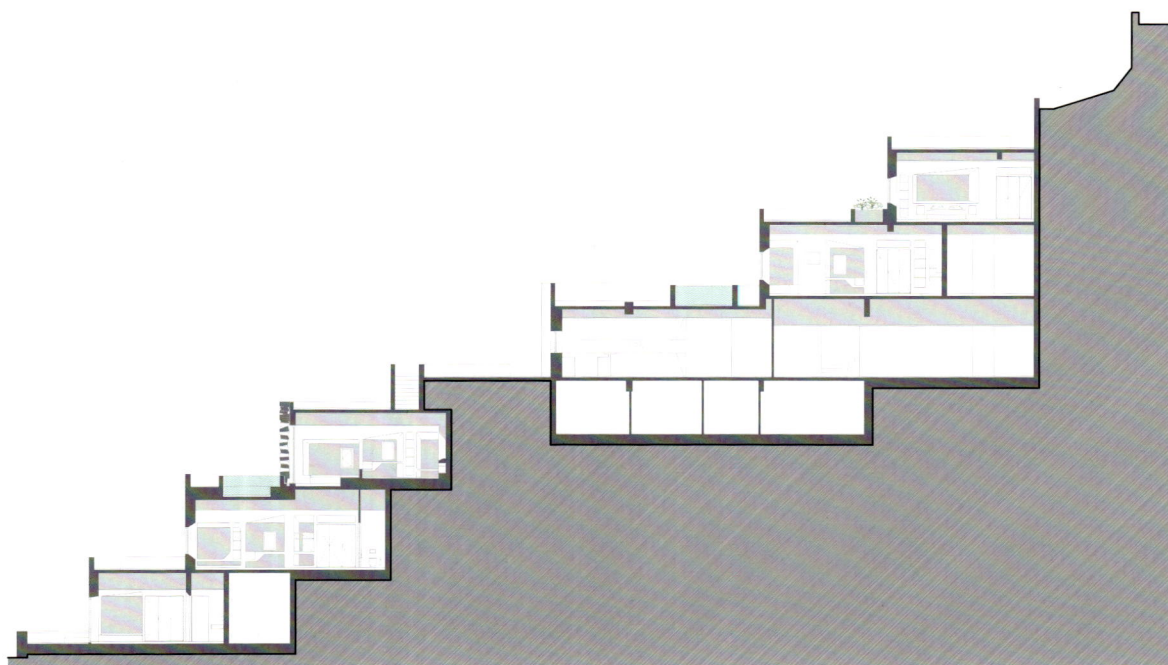

剖面图 Section

The Restaurant
The Bar

外观 2 Exterior II

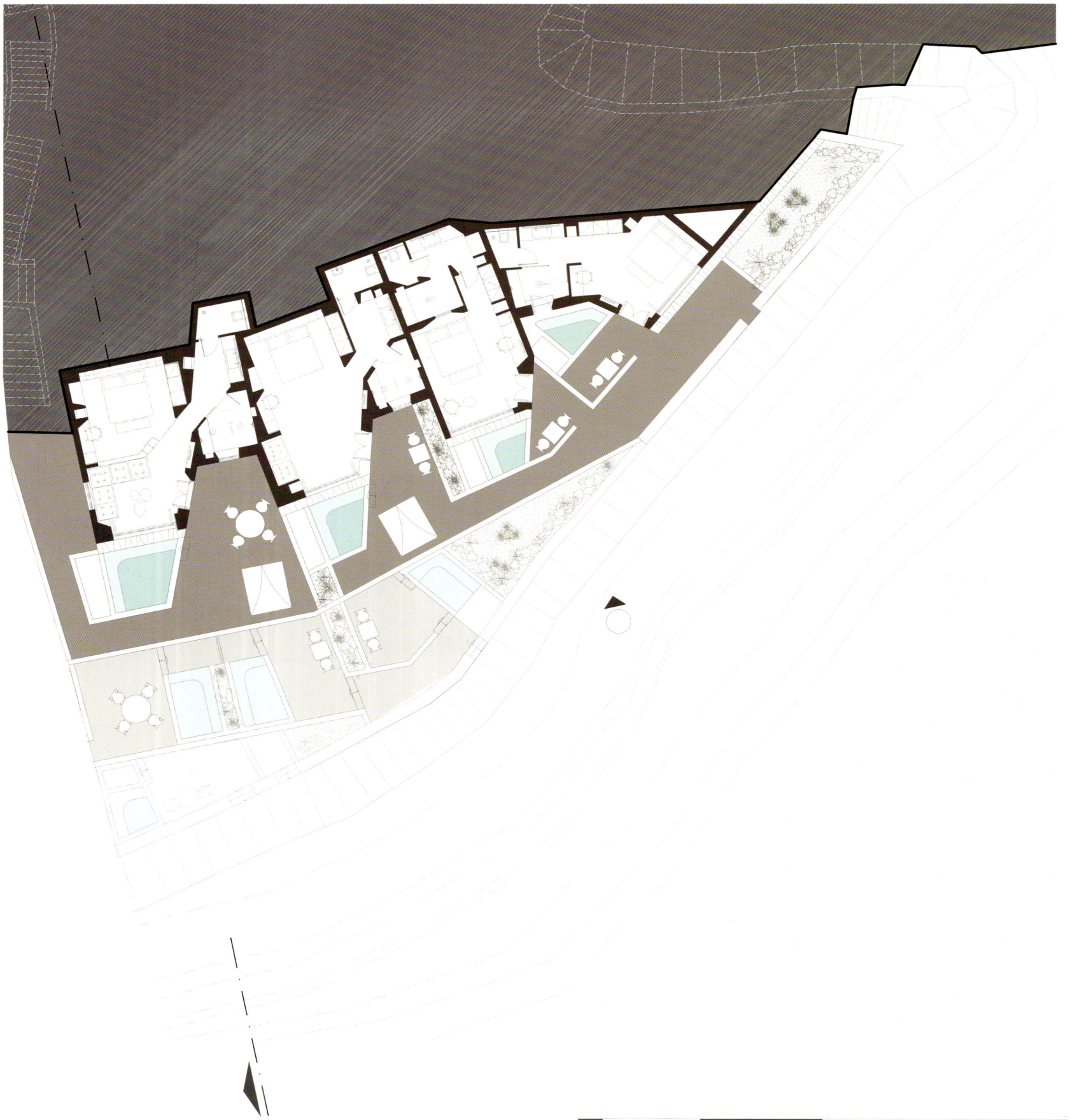

二层平面图 Second floor plan

0　　　　　　　　　　　5　　　　　　　　　10(M)

餐厅 Restaurant

卧室 Bedroom

南港度假屋
South Harbor Resort

地点　Hiroshima, Japan
设计　Suppose Design Office / Makoto Tanijiri
基地面积　1,816m^2
建筑面积　779m^2
总楼面面积　1,242m^2
摄影师　offered by Suppose Design Office

当你第一次看到位于广岛南部辖区的婚礼大堂时，你就不会再考虑去其他地点举行婚礼仪式了。这里有大量的美景：它面向大海，仅有一条马路之隔，另一侧由葱郁的森林包围着。婚礼大堂本身是一个经过装饰的建筑造型，但与周围的自然风景相比显得很适中。在我们的项目中，即使殿堂也不会比自然美景更夺目。该大堂会包含婚礼所需的一切元素，包括门厅、教堂和宴会厅。借助于独立的屋顶和地面设计，宾客们可以在布满鲜花与绿叶的屋顶下感觉室内与外部的空间，同时也可以欣赏四周优美的景色。目前，尽管这个婚礼大堂仅处于规划阶段，但设计师已充分考虑到了其中最微小的细节，决不会影响它的质量。即使是在初期阶段，设计师也已经着手规划建筑平面设计图。从家具到照明设备，再到配件，所有的这一切都会与优美的环境相匹配，体现它的格调。设计师会努力地创造出适合仪式的环境，让人们感受到关怀与重视。"第一次"，它的魅力在于承载了很多希望，同时也有担忧。就像一座山脉，只有爬到山顶的人才能看得到风景，在这里，这种情感得以体现。从这个角度上，设计师相信可以在这个将要建造的建筑中唤起人们相同的感觉。因此，在设计师构思婚礼大堂的最终方案时，会包含一些令人难忘的感受：希望再次见面，希望探索新的优美环境，以及对"第一次"的难忘。

From the first time that you see the Marriage Hall, you will want no other place for your ceremony. The site is in the south of the city of Hiroshima, in the prefecture of the same name. It is a place of abundant natural beauty; the site faces the sea, separated only by a road, and is surrounded on the opposite side by verdant natural forests. The Marriage Hall itself will project an image that is decorative, but the architecture will appear modest when compared with the natural surroundings. In our plans, even the Hall will not outshine the beauty of nature. The Hall will contain everything that might be required elements for a marriage ceremony, including a foyer, chapel, and banquet hall. Thanks to the separate roofs and floor plans, guests will be able to experience for themselves both the interior rooms and the exterior spaces beneath a roof of flowers and greenery, and to appreciate the beautiful surroundings as they move from area to area. At present, even though the Hall is only in the planning stages, we have given even the smallest details our full attention, without compromising on quality. Even in these early stages, we have put thought into designing floor plans and architecture which reflects the location. From furniture to lighting to accessories, everything will suit the beautiful setting with a note of style. We have striven to create a setting appropriate for the ceremonies that it will be hosting, and so we are giving these highlights the care and attention that they deserve. In "For The First Time", the charm lies hidden between a great many hopes and anxieties. Like a mountain, only those who climb can see the view, and in such places a juxtaposition of such abundance and emotion can be found. We believe that from this point on, we can evoke the same feelings in the place that we will create. Thus, when we consider our final plans for the Marriage Hall, we will include the unforgettable feelings of wanting to meet again, of wanting to explore a new and beautiful place, of "For The First Time".

入口 Entrance

外观 Exterior

入口大厅 Entrance hall

室內 1 Interior 1

从门厅看 1 View from foyer 1

南立面 South elevation

室内 2 Interior II

露台 Terrace

北立面 North elevation

从门厅看2 View from foyer Ⅱ

家庭接待室 Family anteroom

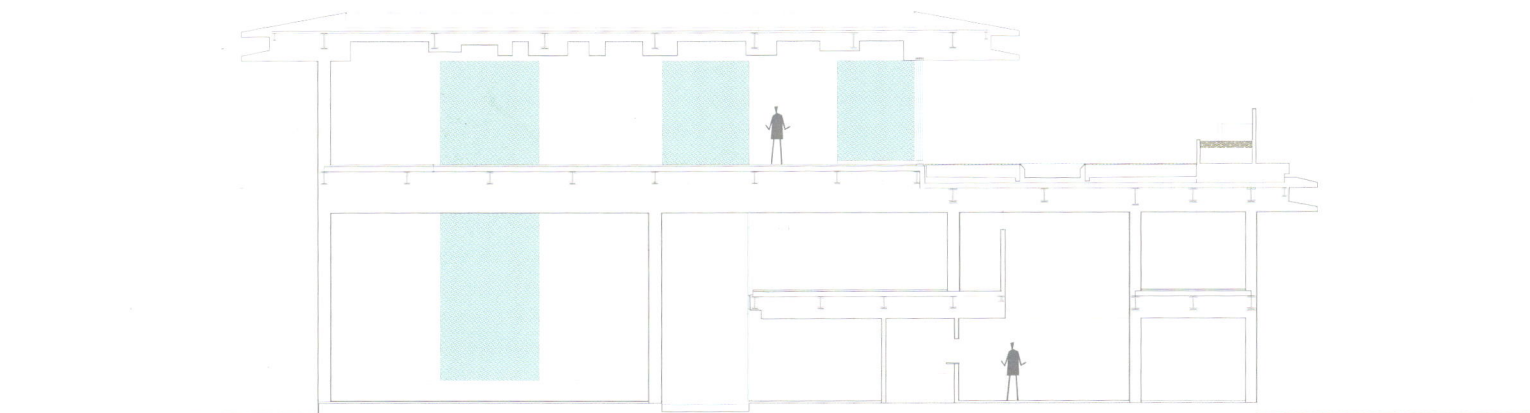

剖面图 Section

小礼拜堂 1 Chapel I

一层平面图 First floor plan

宴会厅 Banquet hall

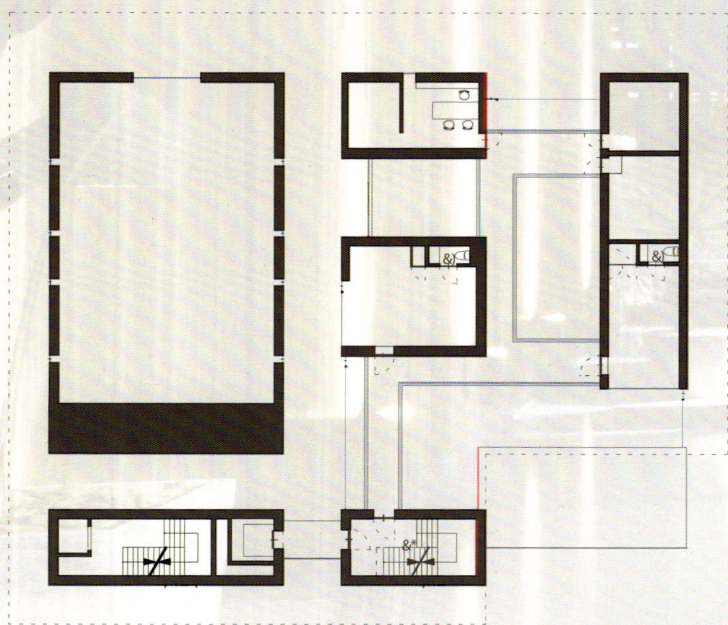
二层平面图 Second floor plan

楼梯 Staircase

三层平面图 Third floor plan

Daemyung Sonofelice

度假俱乐部

Daemyung Sonofelice Resort Club House

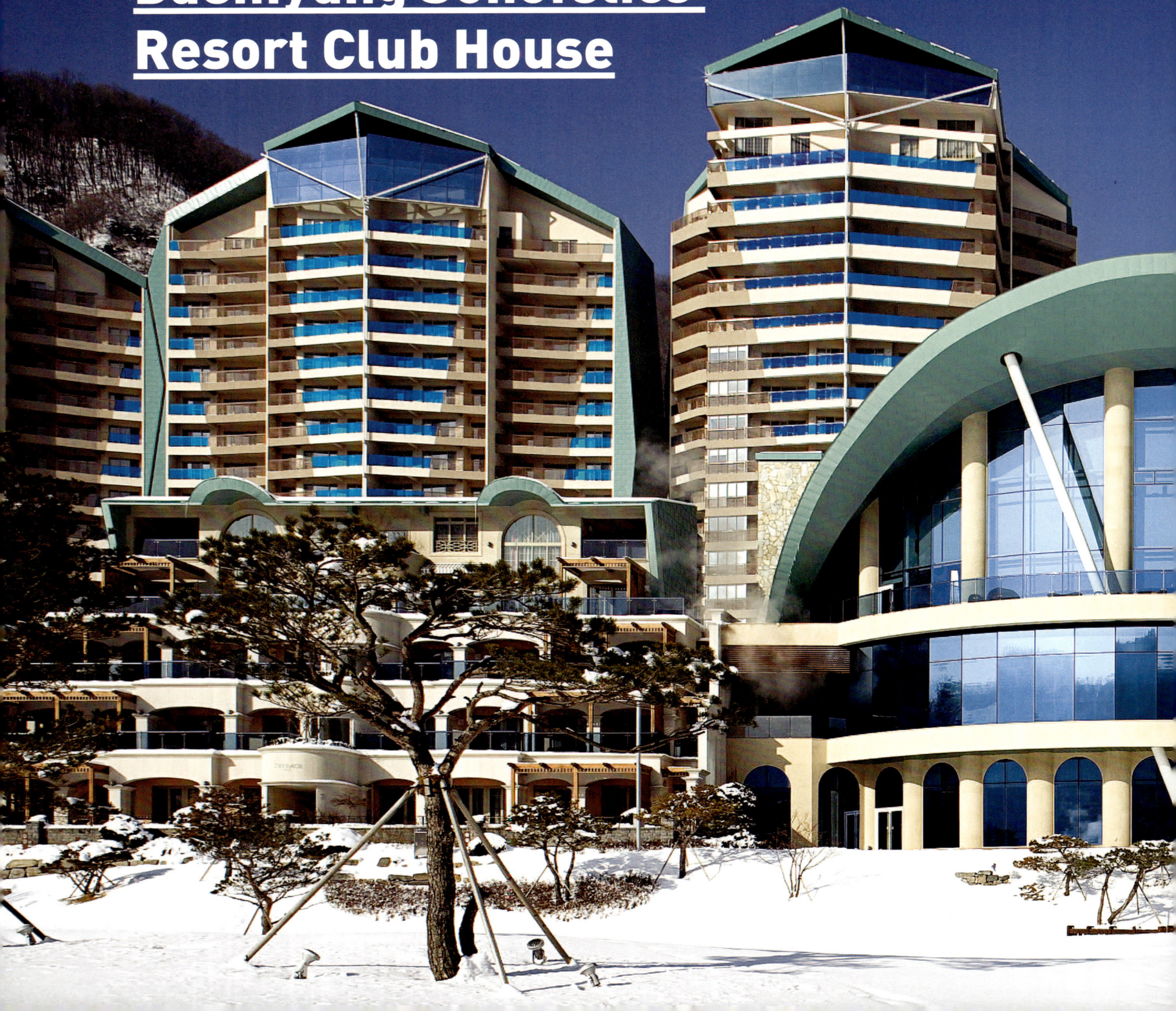

地点　Gangwon-do, Korea

设计　David - Pierre Jalicon

建筑面积　3,770m^2

饰面材料　Wall_Multi-Ceramic Coating Paint, Color Tempered Glass
Ceiling_Green Blue Oxidize Copper Plate
Floor_Carpet, Wood, Wall_Artificial Grass, Printed Glass
Ceiling_Printed Glass, Veneer(int.)

施工单位　DAEMYUNG CONSTRUCTION CO., LTD.

摄影师　Kim Jae-yun

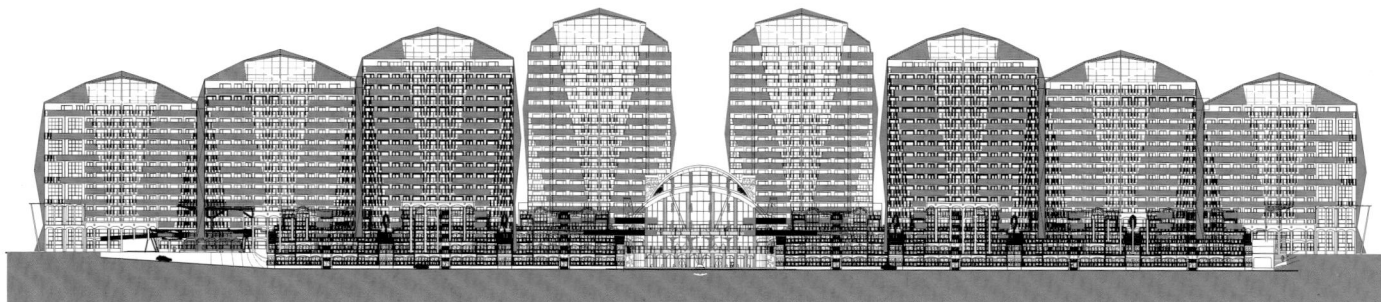

立面图 A Elevation A

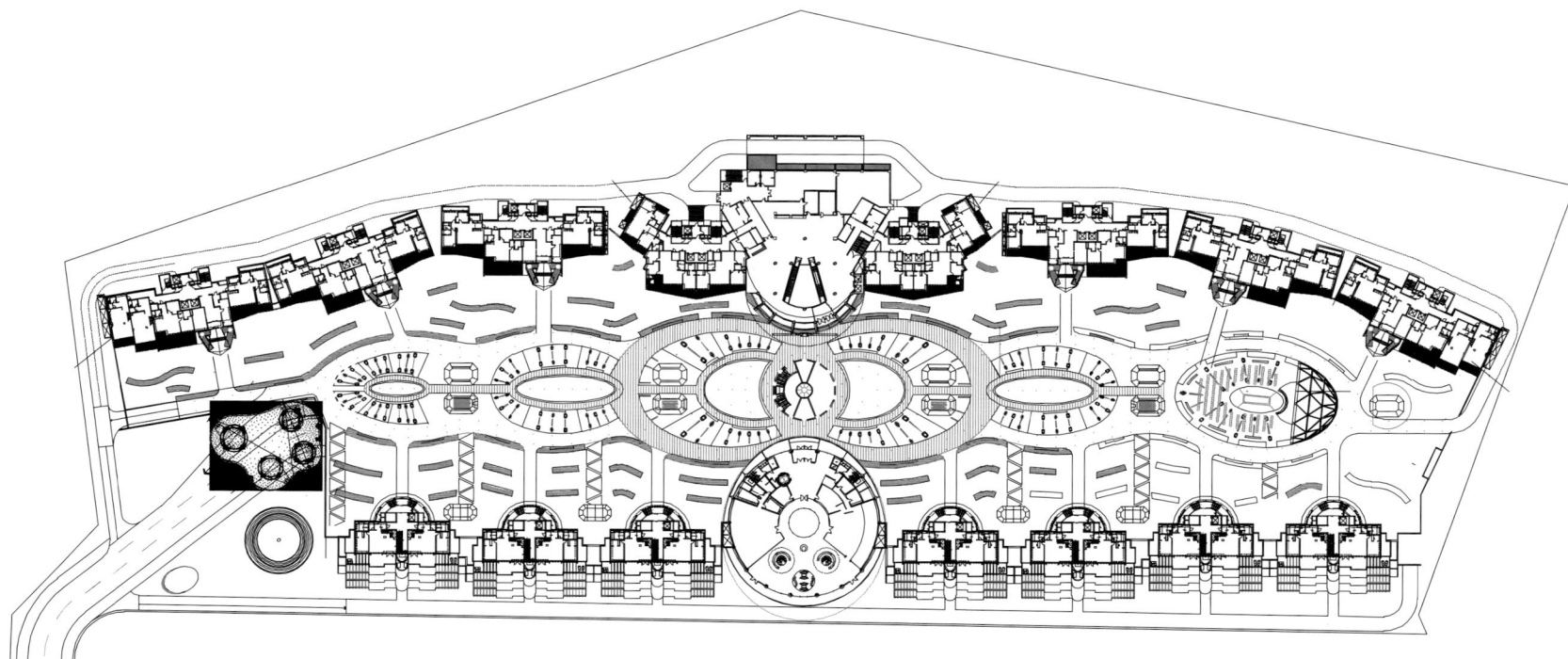

楼层平面图 Floor plan

入口 Entrance

CONCEPT 1

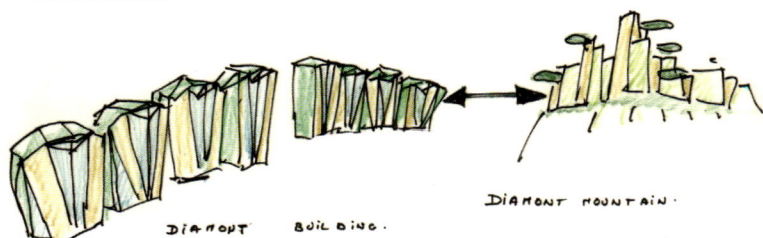

DIAMONT   BUILDING.

DIAMONT MOUNTAIN.

CONCEPT 2

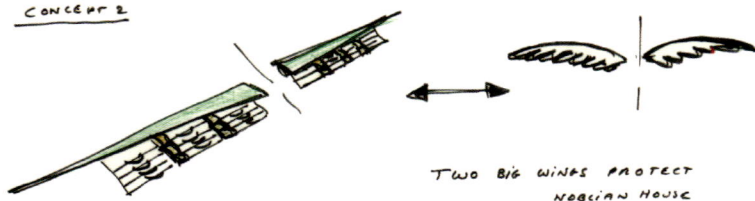

TWO BIG WINGS PROTECT
NOBLIAN HOUSE

CONCEPT 3

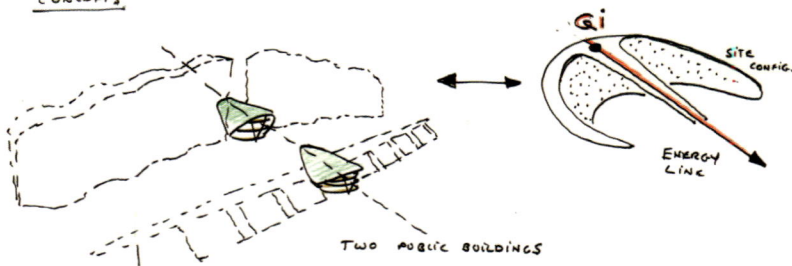

TWO PUBLIC BUILDINGS

CONCEPT 2 : JEWELRY SYMBOL OF LUXURY

CONCEPT 3 : 1. JEWEL PROTECTED BY BOWL SITE
A BOWL SITE HIGHLIGHTED BY A SPARKLING JEWEL

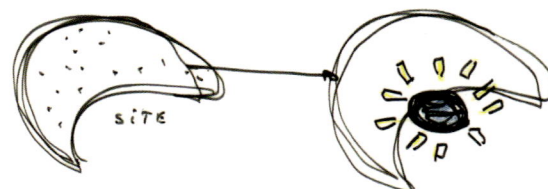

SITE

2. AN ARCHITECTURAL STYLE WHICH DO RESPECT THE SITE

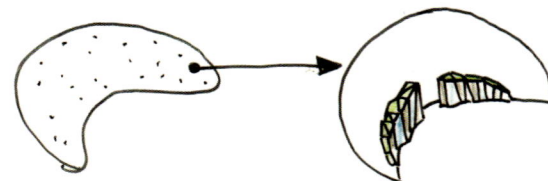

3. DIA - MOUNT RESORT : HARMONY OF LUXURY AND NATURE

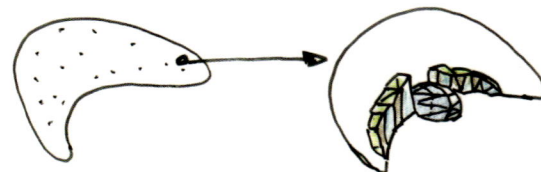

概念草图 Concept sketch

Sonofelice 俱乐部的设计目的是保持自然的状态，它运用了人们肉眼能看到的最微小的元素，比如细胞、气泡和光纤。每个楼层都有不同的主题：一层的"绿色细胞"由大厅和休息厅组成，地下一层的"平稳波浪"设有桑拿房和健身房，地下二层的"水滴"由游泳池和水疗中心组成。

当你行走在水平路面上，迎接你的是一个拥有大型门厅和休息厅的主厅，其视野非常开阔。为了使它的建筑概念更加清晰，所有的墙壁都采用照明和细胞单元模式，这样墙壁看起来像是一个浮动的绿色细胞。这种处理方式会给游客们带来一种全新的体验，让他们感到无论走到哪里，都被细胞墙包围着。地面像一个巨大的光纤，从建筑的一端穿透到另一端，形成一个充满生机的空间。在地下一层，重新出现的细胞墙给人一种在空间上相互连接的感觉，再加上厅内半透明的穹顶天花板和吊灯所散发出的柔和的波浪效果，能够使游客在这里得到放松与休息。地下二层的天花板由色彩斑斓的、不规则的水滴装饰而成，有点像凝结的水蒸气，这种视觉设计能给人们增加娱乐效果。在俱乐部中心，有一个穿透三层楼的巨大的玻璃柱，使每一层都有光线。这里呈现出有活力的自然景象，游客们能发现共存的意义并且感受到内部与外部的联系。

Based on the motive of natural shapes, Sonofelice Club was built upon the smallest elements you can see with your bare eyes such as a cell, a bubble and a fiber. Each floor has different theme: "a green cell" for the 1st floor occupied by a lobby and a lounge, "a smooth wave" for the 1st basement floor where a sauna and a fitness room are located, and "a drop of water" for the 2nd basement floor occupied by a swimming pool and spa.

Upon walking onto the ground level, you are welcomed by the main hall with a large lobby and a lounge with an open view. To emphasize the concept more clearly, all the walls are applied with the lighting and cellular pattern so that they seem as a floating green cell. Such a solution makes the visitors feel as if they were enclosed by cells wherever they go, creating a new experience. The floor looks like a huge fiber penetrating the building from one end to the other to produce a dynamic and vital space. On the 1st basement floor, the cell wall of the 1st floor reappears to give a sense of spatial connection as well as underlining the purpose of the building as a relaxing lounge through a gentle wave of a translucent dome ceiling and the chandelier in the lobby. The 2nd basement floor provides amusement through the ceiling decorated by colorful and irregular water drops so that it seems to be condensed vapor. In the center of the club, there is a huge glazed column that penetrates three floors and delivers light to each of them. In this place that presents a dynamic scene of nature, the visitors find meanings of coexistence and feel the connection between the inside and outside.

餐厅 & 咖啡厅 Restaurant & Cafe

大堂 Hall

图书室 Library

休息室 Lounge

健身房 1 Fitness 1

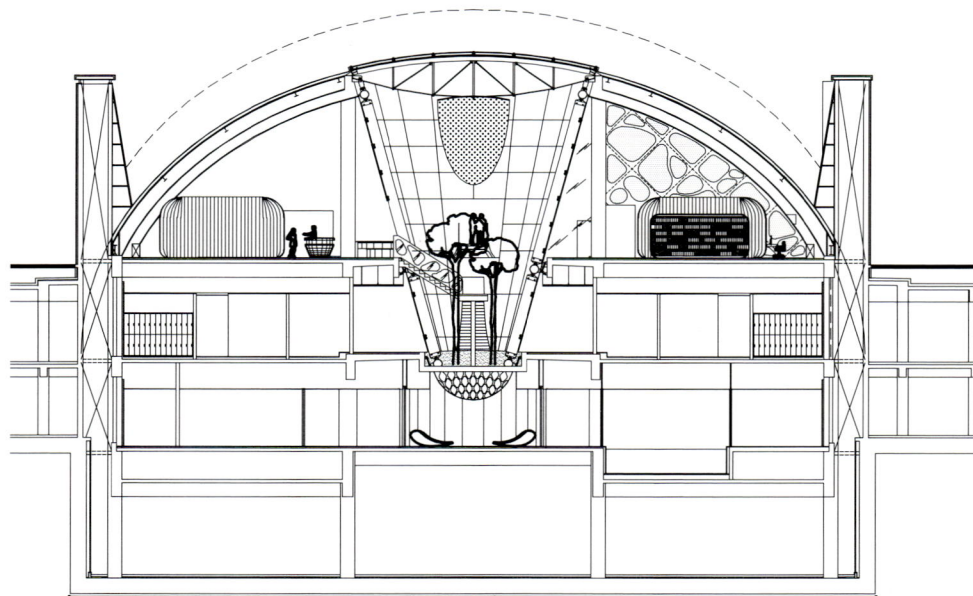

水平剖面 Horizontal section

水疗中心 Spa

桑拿房 Sauna

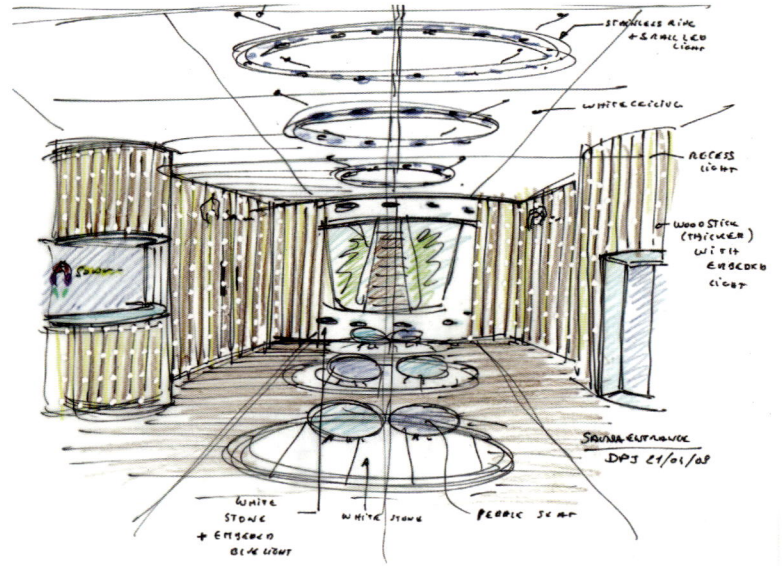

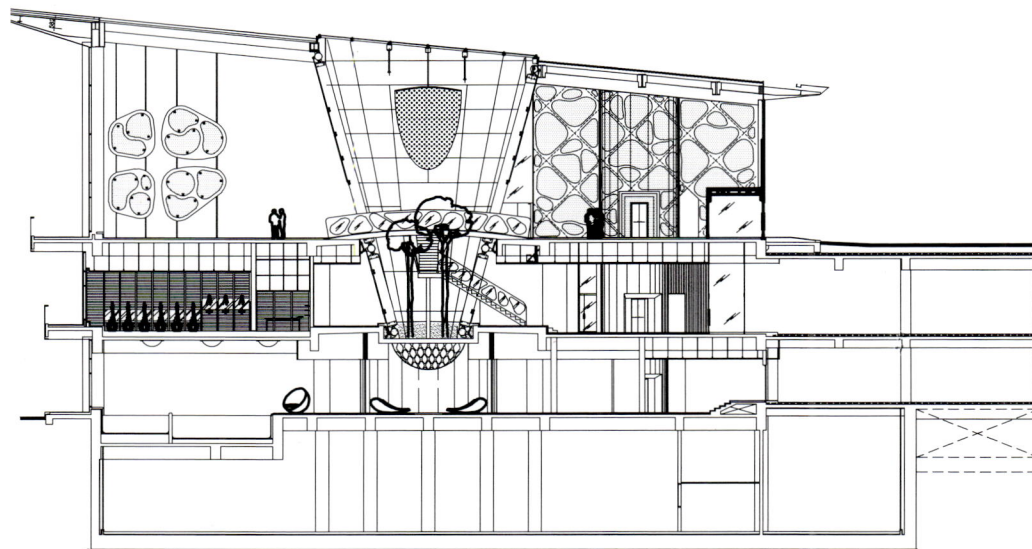

垂直剖面 Vertical section

音乐治疗室 Music therapy room

BARRISOL CEILING

GLASS BUBLE.

BARRISOL BUBLE

DROPING BEAD IN FRONT OF ACRYL TUBE.

WOOD DECK.

STAINLESS BUBLE.

GREEN BUBLE

PRINTED TILES.

1. 芳香治疗室
2. 音乐治疗室
3. 酒吧
4. 游泳池
5. 治疗室
6. 大厅 I
7. 接待处
8. 桑拿房入口
9. 健身房
10. 平台

11. 淋浴区
12. 浴室
13. 桑拿房
14. 信息台
15. GX 室
16. 连接桥
17. 图书室
18. 酒吧
19. 休息室
20. 餐厅 & 咖啡厅

1. Aroma therapy
2. Music therapy
3. Bar
4. Pool
5. Therapy room
6. Hall
7. Reception
8. Entrance of sauna
9. Fitness
10. Dock

11. Shower area
12. Bath
13. Sauna
14. Information desk
15. GX room
16. Bridge
17. Library
18. Wine bar
19. Lounge
20. Restaurant & Cafe

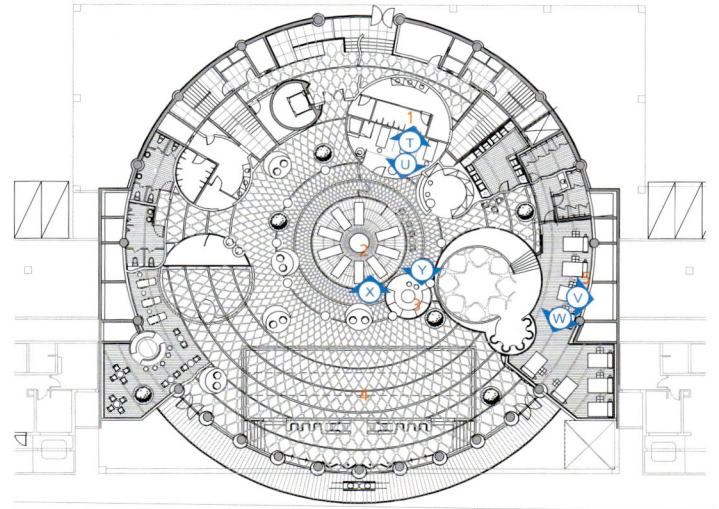

二层平面图 Second floor plan

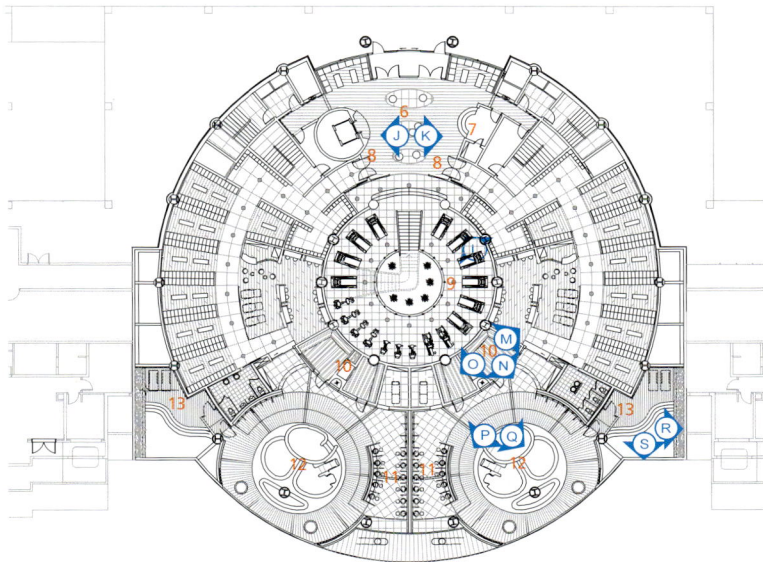

地下一层平面图 Basement first floor plan

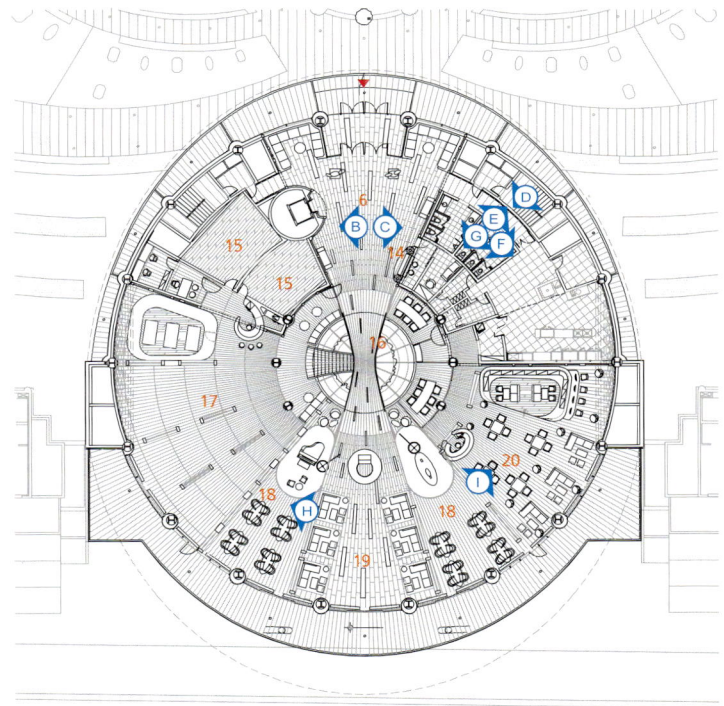

一层平面图 First floor plan

韩华金海湾高尔夫度假村
Hanwha Golden Bay
Golf & Resort

地点　Chungcheongnamdo, Korea

设计　GANSAM Architects & Partners

基地面积　1,385,535㎡(golf club) + 7,313㎡(condominium)

建筑面积　5,184㎡(golf club) + 2,031㎡(condominium)

总楼面面积　12,446㎡(golf club) + 8,391㎡(condominium)

楼层　2FL(golf club), 6FL(condominium)

结构　Reinforced concrete, Steel framed reinforced concrete

饰面材料　Stucco, Rustic stone, T24 Transparent pair glass, Roof tile(ext.)

结构工程师　HANWHA Engineering & Construction

摄影师　Park Young-chae, Yoon Jea-hyuk

N

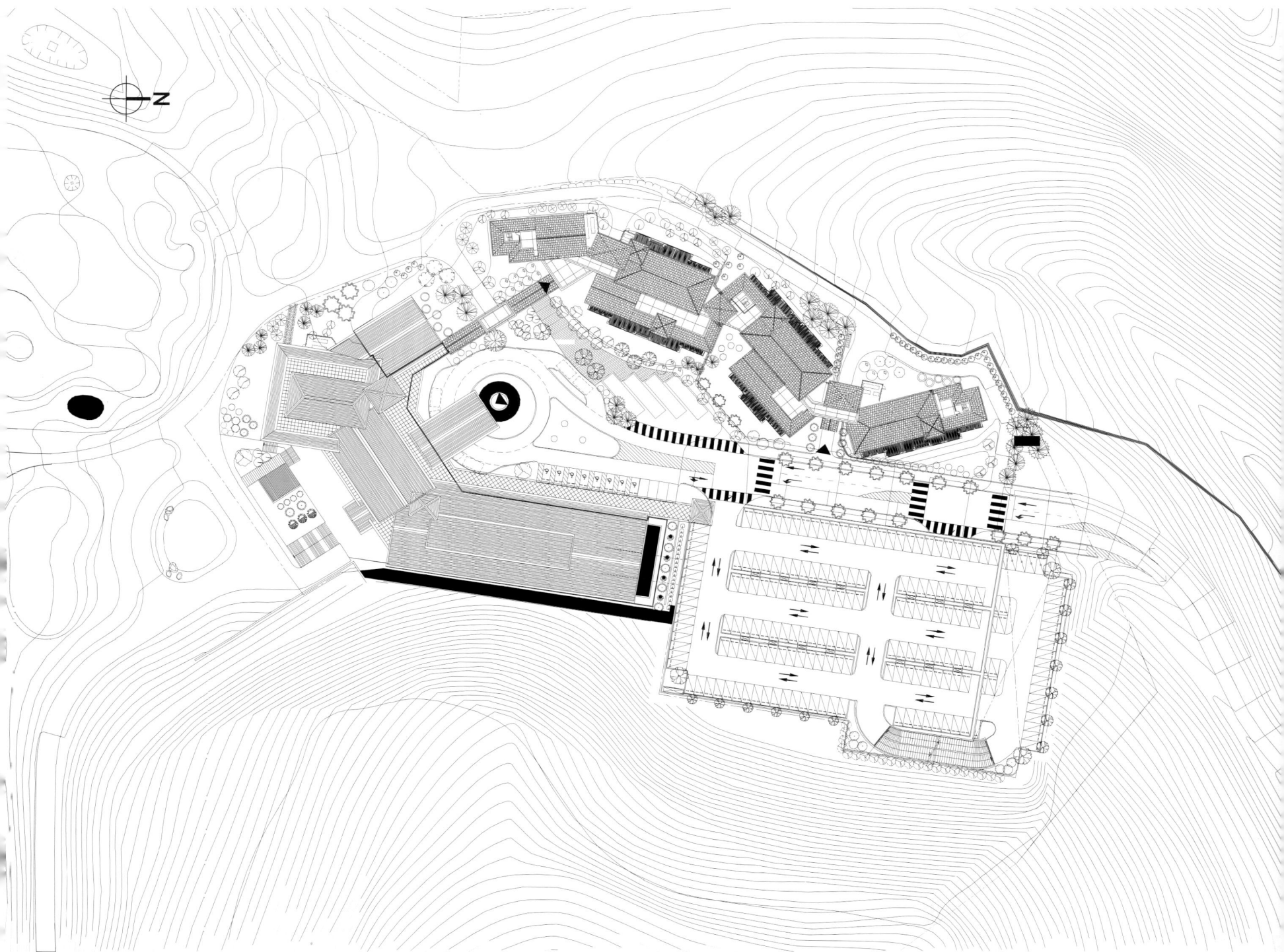

总平面图 Site plan

全景 Overall view

韩华金海湾高尔夫度假村，来自大自然的礼物

梦想的高尔夫度假村终于在这片神奇的土地上开业了，这里大海、海岸、山谷和山脉和谐地融合在一起。它的名字叫"金海湾"，寓意着"金色满满的海湾"。拥有八大景观的大安解颜国家公园在韩国非常有名，金海湾高尔夫度假村就位于这座公园内。由高尔夫皇后阿尼卡·索伦斯坦女士设计的高尔夫球场也很出名。除了韩国的八大景观，大安还拥有 32 个海滩和将近 100 个美丽的岛屿。其中，Sindoo 是韩国最大的海滨沙丘，Sindoo 沙丘被指定为第 431 号自然保护区，是生态旅游的最佳去处。金海湾高尔夫度假村位于连接 Baekhwa 山脉与 Anmyeon 岛的海岸上。

度假村有着 2 300 000m² 的广阔场地，拥有一个 27 洞的高尔夫球场和 56 个海滨套房。整体上，度假村延续意大利的托斯卡纳风格。大安与托斯卡纳一样因其得天独厚的自然环境而闻名。石块和木材是直接从意大利空运过来的。褪色的黄色石块、水泥墙和红砖屋顶，使这里的场景看起来更加舒服。这里有 56 个"兼高尔夫功能的酒店"，每个套房有 140m²，并配有两个主卧室，这样使每个房间都有完善的隐私空间。房间的内部装潢也运用了柔和的托斯卡纳风格，在这个会所里，可以将高尔夫球场的美景尽收眼底。运用老式手法设计的环境让人觉得很舒适。外墙饰面采用了石材、灰泥和木材，从而使其与自然和谐统一。托斯卡纳村毗邻会所。这个 7 层的建筑拥有 56 间套房，当高尔夫爱好者们透过窗户欣赏西海金色海浪的时候，可以感受到一种异国情调。

The golf club with gift from nature, Hanwha Golden Bay Resort

At last, a dream golf resort has opened in the blessed land where the sea, coast, valley and mountain are in harmony. The name is "Golden Bay", meaning the "bay with golden color". The Taean haean National Park is famous with the eight great sceneries in Taehan, and Golden Bay Golf & Resort is in that park. The golf course is also famous for having been designed by the golf empress, Ms. Anika Sorenstam. In addition to the eight great sceneries in Taehan, Taean has 32 beaches and more than 100 beautiful islands. Among them, Sindoo Dune is the largest seaside sand dune in Korea. Designated as No. 431natural monument, Sindoo Dune is the best place for ecological tourism. Golden Bay Golf & Resort is located at the seacoast connecting Baekhwa Mountain and Anmyeon.

The resort with expansive site of 2.3 million m², has 27 holes golf course and 56 suites seaside golf village. As a whole, the resort consists of Tuscany style spaces. Taean is no less famous than "Tuscany" with its gifted natural environment. Stones and woods, were directly air-transported from Italy. One becomes more comfortable by looking at the stones and plastered walls with faded yellow color and red brick roofs. There are 56 "golftels" (golf+hotel) and each 140 m² suite has two master rooms so that each room can have perfect privacy. The internal decoration of the rooms also has bland Tuscany style design. The clubhouse has a great view of the golf course. It has a cozy atmosphere while designed in old fashioned way. External finishing used stone, plaster and wood so that it would be in harmony with the nature. Tuscany Village is located next to the clubhouse. The 7-story building has 56 suites. Golfers can enjoy an exotic mood while looking at the golden waves of Western Sea out of window.

外观 2 Exterior II

公寓 Condominium

前立面 1 Front elevation I

外观 3 Exterior Ⅲ

高尔夫俱乐部 Golf club

前立面 2 Front elevation Ⅱ

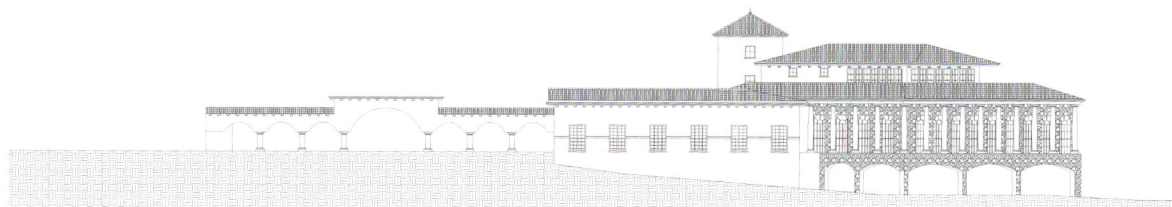

西南立面 Southwest elevation

入口 Entrance

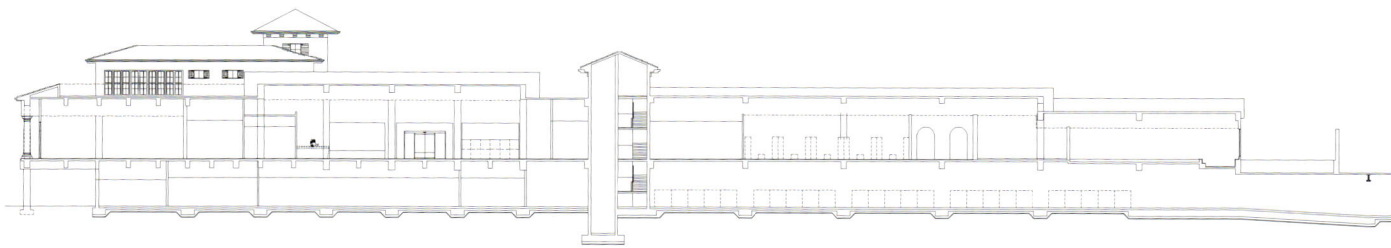

横剖面 Cross section

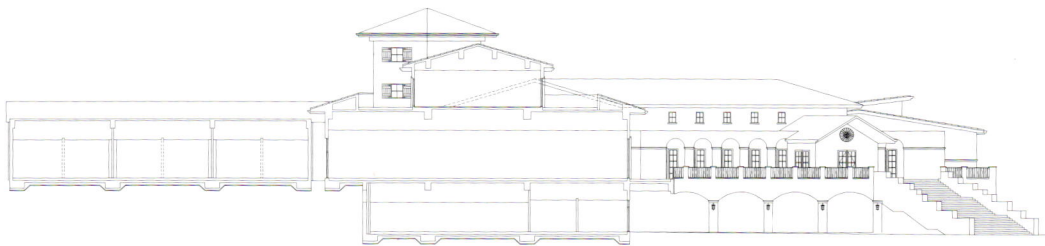

纵剖面 Longitudinal section

餐厅 Dining room

房间 Room

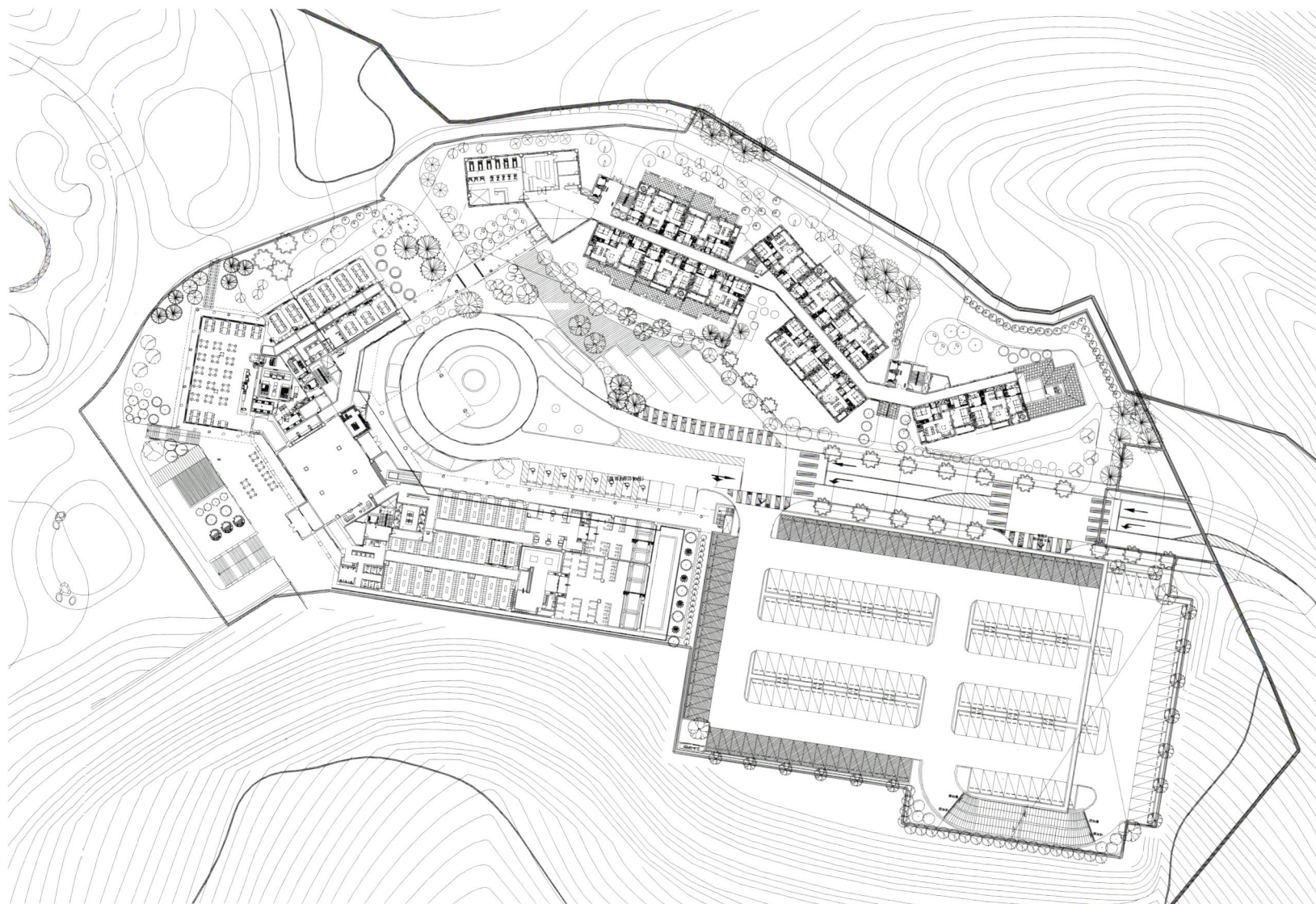
一层平面图（高尔夫俱乐部）＋二层平面图（公寓） First floor plan(golf club) + Second floor plan(condominium)

餐厅 Restaurant

Saffire 度假村
Saffire Resort

地点　Coles bay, Australia
设计　Circa Architecture / Circa Morris-Nunn Walker
参与设计人员　Peter Walker, Poppy Taylor, Jarrod Hughes,
　　　　　　Robert Morris-Nunn, Ganche Chuam,
　　　　　　Judi Davis, Chris Roberts, Gary Fleming, Tina Curtis
甲方　The Federal Group
室内设计　Chhada Simbiada Interior Design
景观设计　Inspiring Place
结构工程师　Gandy, Roberts
建筑工程师　Fairbrother Construction
摄影师　Declared in picture

N

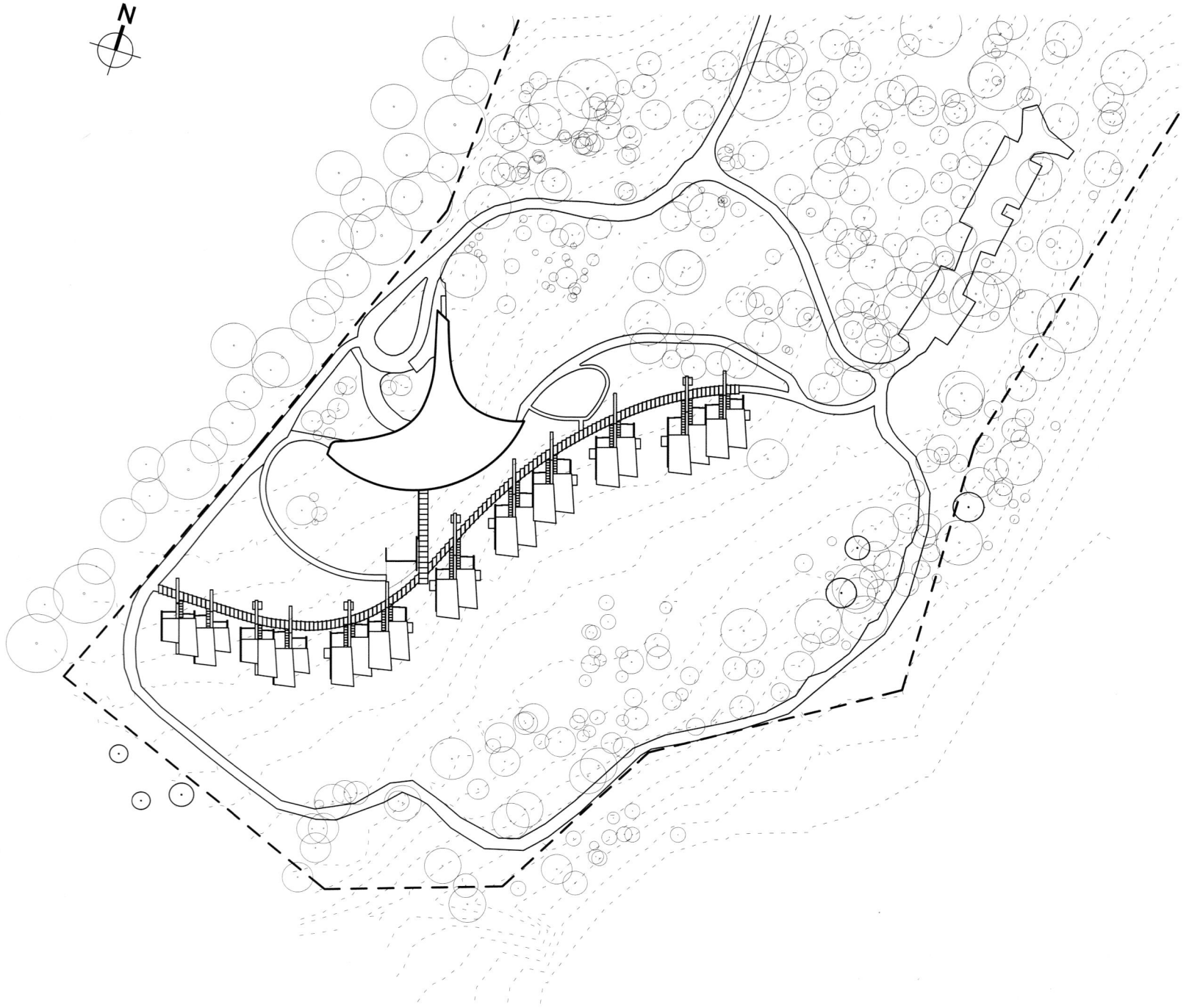

总平面图 Site plan

鸟瞰图 Bird eye view

外观 1 Exterior I

© Peter Whyte

设计陈述

从一开始，Saffire 就被想象成是一个重新定义塔斯曼尼亚岛旅游业的图标性项目。当我们继承了这一点之后，这处景点就不再是以前那个废弃的公园，所以这个项目就变成了关于修缮这个场所并诠释它独一无二的品质，同时创造一个可以给人带来超凡体验的空间。带着这种想法，设计师把主要建筑放在旅途的终点。在那里，Hazards 都是被保护和展现出来的。最后作为一个目的地，它呈现出了澳洲大蚝湾的全景。这个度假胜地还与地点有机结合在一起。它的形态唤起了人们对海滨、陆地、沙丘、波浪、海洋生物的记忆。现在的旅途就是一个从自然保护区到套房中个人空间的过程。这里的小波浪和生物就好像在标记着潮汐和海岸线。各套房之间的走廊就仿佛是一片海滩，而公寓就像是停泊在沙滩上的船只。

每间套房都是封闭、私密的，在比例上更具私密性，但是却拥有个性化的视野。在甲方和其他建筑师探讨这个项目的时候，对于项目的方向有了改变，并要求我们对其进行重新设计，而此时我们已经做了别的酒店项目，并且和甲方也建立了很好的关系。其地点位于塔斯马尼亚岛的东海岸，能够俯视整个大蚝湾、Hazard 和 Freycinet 半岛。Hazards 是一个由粉色花岗岩形成的独一无二的地质形态，让它看起来粉红中略带橙色。这个项目就坐落在这个天然的海滨景观之内。起初的设想是建造 150 个房间，但是应甲方的要求重新考虑，最后变成一个更小规模、更具私密性的度假胜地，仅拥有 20 间套房。这是我们一致同意的决定。我们的甲方联邦集团，已经在整个塔斯马尼亚岛主要的旅游景点拥有一系列酒店地产（如 Hobart、Cradle Mt、Strahan、Port Arthur 等等）。该度假胜地通过在 Freycinet 为游客提供高端的旅游住宿环境，以满足四星级的标准。新度假村意在成为人们旅程的终点，并迎合当地、洲际或国际游客的品位。

© Peter Whyte

外观 2 Exterior II

持续性

这个地点在施工以前就已经脱离了先前的用途，我们把这个项目看作是对于周围景色的"复原"。通过和景观设计师（Inspiring Place 公司）的共同合作，建筑的地点位于完全保留了原有植被和树木的地方。在施工过程中，设计师就建造了保护区，并将大片的再种植物作为景观设计的一部分，还将与可持续设计相关的、可能出现的问题都考虑在内，也权衡了建筑场址较远这一现实问题。对于项目来说，首先应该考虑用水问题，尤其是在受干旱影响的地区。和污水处理系统一样，作为项目的一部分，新的雨水收集和贮藏设施已经建好。滴落在屋顶上的雨水被收集起来，并重新利用于水池以及高效的节水设施中。由于该地以凉爽的气候为主以及向南的朝向，另一个需要考虑的重要因素是该景点的保温功能。所有的建筑都具有良好的隔热功能，并且安装了高性能的玻璃窗。高效的热水器和空调系统也都投入使用。

规划

这个规划基本分为三部分。首先，客人们在到达主要接待处（sanctuary）之后，通过一条像通道一样的长"码头"进入到一个观赏平台，在这能够俯视港湾。人们可以从这个更高的平面下到餐厅，在这也能看到全景。该接待楼的最底层设有 SPA（跟公共区域是分开的）、健身房、董事会会议室和画廊。这里还与客房相连。第二，20 间客房设置在一条蜿蜒的通道上，每一间都经过精心设计，以确保置身其中的人们能将 Hazard 的景观一览无余，并且为客人提供一个私人空间。这里共有三种套房类型，位于景点的西面，距离接待处最远。每间套房中都配有一个朝向景观的甲板，以及一个位于北侧的私人庭院，可以从通道进入。第三，分布在通道中以及在接待处和套房之间的是住宅设施，如储藏室、房屋管理、仓库，它们是为套房提供服务的，但却离景点很远。

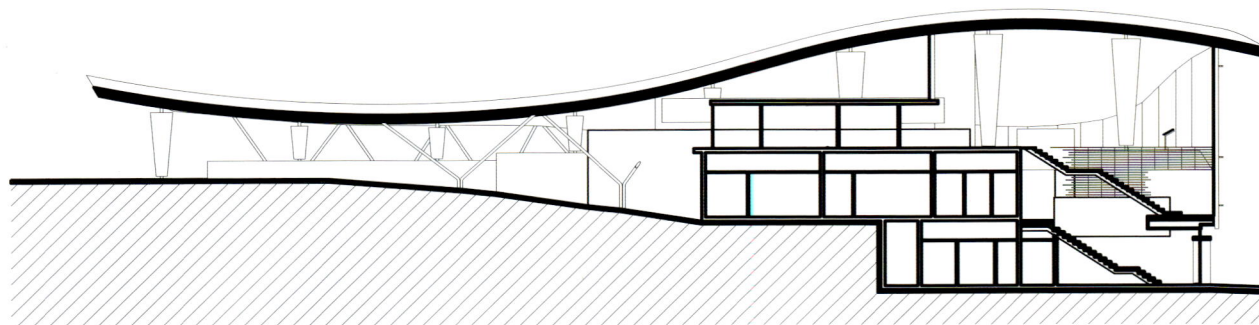

剖面图 Section

Design Statement

From its inception, Saffire was imagined as an iconic project to redefine tourism in Tasmania. The location, when we inherited it, was scarred from its previous use as a disused caravan park, so the project became as much about repairing the site and interpreting its unique qualities as it was about creating a space from which it could be experienced. With this in mind, we shaped the main building as the end point in a journey, in which views of the Hazards are shielded and revealed and finally presented as a destination which is a panoramic overview of Great Oyster Bay. The resort is also organic in its relationship to the site. Its form evokes memories of coastal land forms, dunes, waves or sea creatures. The journey now is one that moves from the monumental to the personal space of the suites. These are small waves or creatures, arranged on the site as if marking the tidal shoreline. The passage between the units is a metaphor for a beach, the suites moored like small craft run up onto the sand.

Each suite is enclosing and private, yet opens to an individually personalised view − except this time much more intimate in scale. Our client was developing the project with other architects, however when a change of direction for the project was sought, we were asked to completely re-design it − we had been working on another hotel project with the client and had formed an excellent working relationship. The site is located at Coles Bay on the east coast of Tasmania, and overlooks Great Oyster Bay, the Hazard and the Freycinet Peninsula. The Hazards are a unique geological formation of pink granite that give it an orange pink tint. The project site is located within, what is extensively a natural native costal landscape. The original brief was for a much larger development of 150 rooms, but this was rethought by our client and eventually became a far smaller scale, more intimate resort, of 20 private suites. This was a decision with which we were in total agreement. Our client, the Federal Group, already own a series of hotel properties at prime tourist destinations throughout Tasmania (Hobart, Cradle Mt, Strahan, Port Arthur, etc). The intent of the resort was to provide high end tourist accommodation at Freycinet to compliment a 4 star lodge already owned by another site.

The new resort is intended to be a destination in its own right and will mainly cater to inbound − interstate or international − guests.

Sustainability

The site, prior to construction, was extensively scared from its previous use, and we saw the project as a way of "healing" the surrounding landscape. In collaboration with the landscape architects (Inspiring Place), the buildings were located to retain all of the existing vegetation and trees. Protection zones were established during the construction period and extensive replanting was undertaken as part of the landscape design. Where possible issues relating to sustainable design were considered however this was also balanced against the practical considerations due to construction on a remote site. A major consideration for the project was water usage in a typically drought affected area. New rain water collection & storage infrastructure were built (off site) as part of the project as well as sewage treatment facilities. Rainwater from roofs is also collected and re-used in the reflection pools, as well as water efficient devices being specified. Another important factor, due to the predominately cool climate and south facing site, was heating the resort. All buildings are well insulated and high performance glazings have been installed. Energy efficient water heating and air conditioning systems are used.

Program

The program is basically divided into three parts. Firstly, the guest arrives at the main reception building (the sanctuary) which is entered through a long "jetty" like walkway to a viewing platform that overlooks the bay. From this upper level, you descend towards the view to the main dining lounge level which also affords panoramic views of the whole site. The lowest level of the reception building accommodates the spa (privately located away from the public areas), gym, board room and gallery. It also provides a link to the guest suites. Secondly, 20 guest suites are located along a serpentine walkway, each carefully sited to capture views of the Hazards and provide a privacy for the guests. There are three suite types with the premium located to the west of the site and furthest from the reception. Each suite has a deck located towards the view and a private courtyard located towards the north which is entered off the walkway. Thirdly, interspersed along the walkway, and located between the suites and the reception are back of house facilities − pantry, house keeping and storage − that service the suites yet kept out of site.

休息室 Lounge

© George Apostolidis

© Peter Whyte

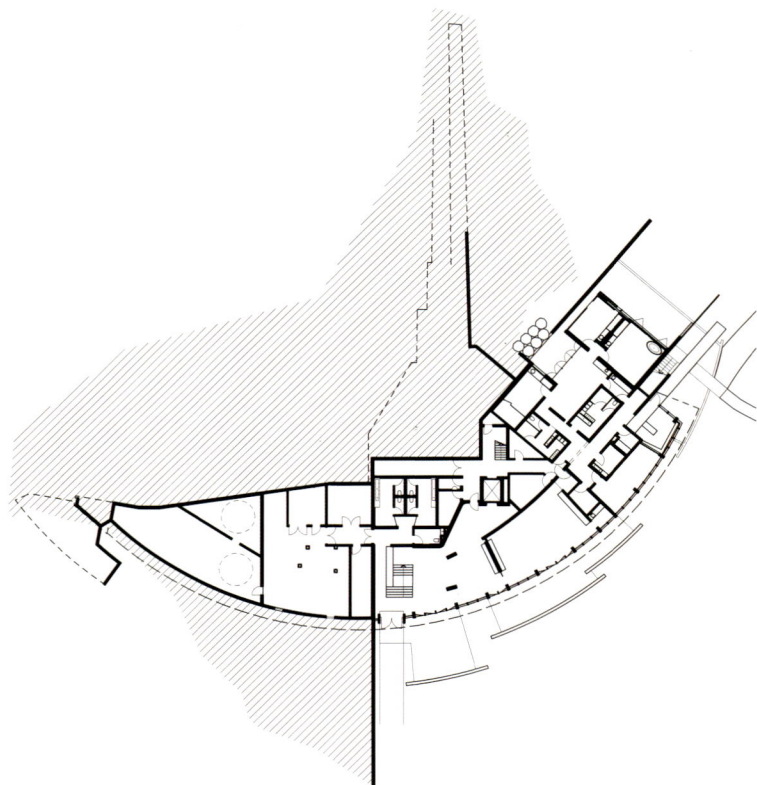

一层平面图 First floor plan

室内 1 Interior

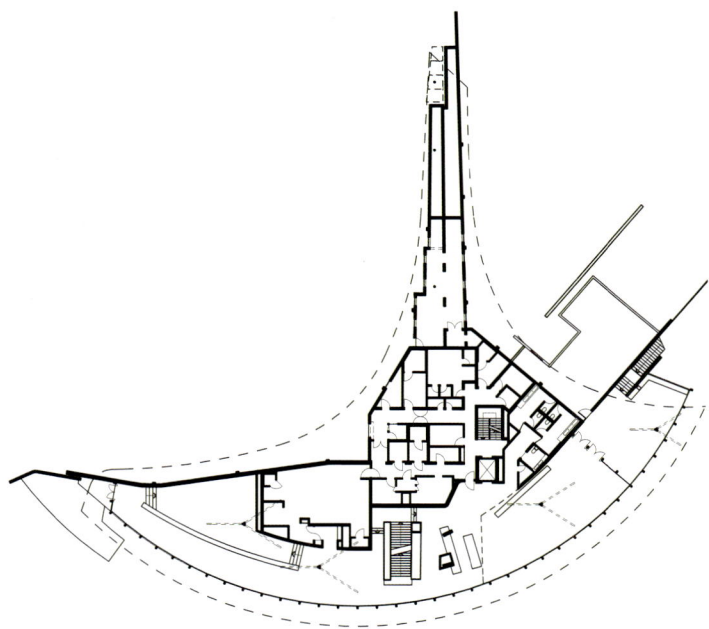

二层平面图 Second floor plan

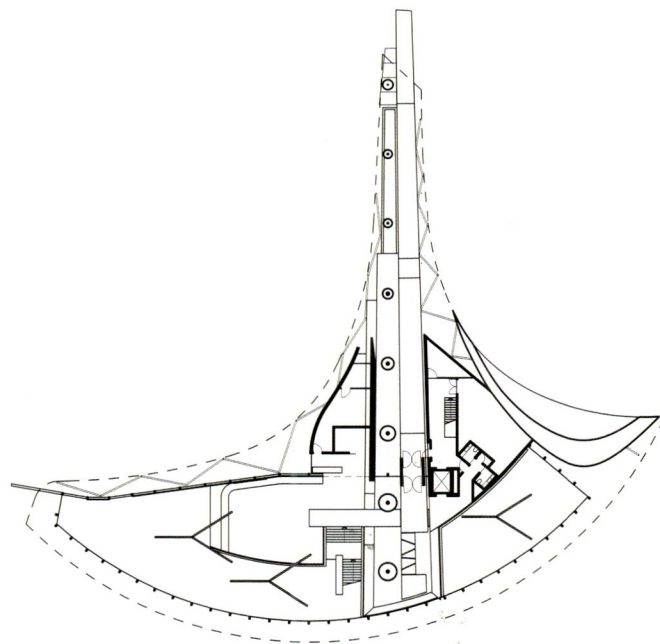

三层平面图 Third floor plan

© Peter Whyte

© Peter Whyte

室内 3 Interior III

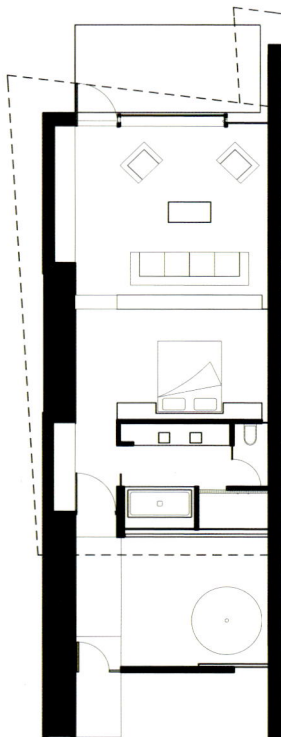

典型平面图——豪华型 Typical floor plan _ Deluxe　　　典型平面图——奢华型 Typical floor plan _ Luxury　　　典型平面图——经济型 Typical floor plan _ Premium

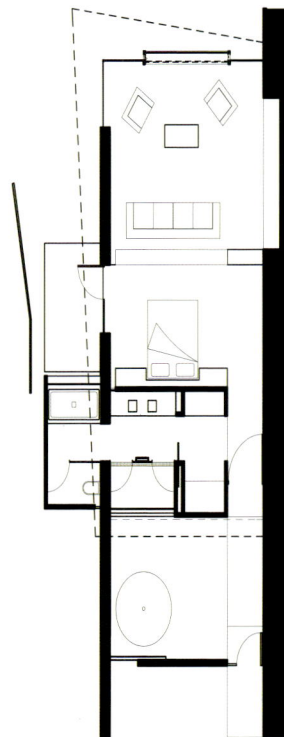

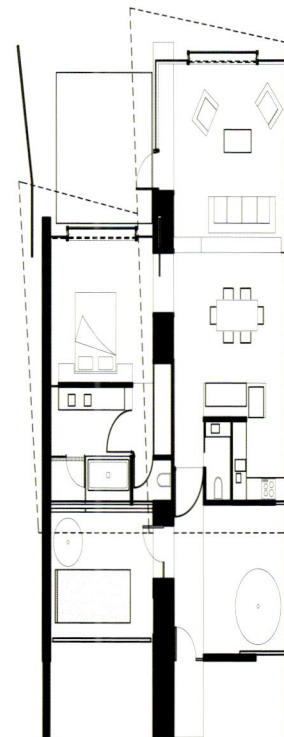

济州岛凤凰度假村 Vella Terrace 公寓
JEJU Phoenix Island Vella Terrace

设计　Gansam Architects & Partners / Kim Tai Jip,
Han Ki Young, Kim Cheon Haeng

地点　Jeju-do, Korea

基地面积　59,078m^2

建筑面积　16,316m^2

总楼面面积　50,925m^2

楼层　B2, 5FL

结构　Reinforced concrete

参与设计人员　Joo Myung Joong, Moon Jong Guk, Jeong Seong Hoon,
Lee Yoon Soo, Ji Hyun Ook,
Lee Jeong Hoon, Kim Hyun Ho, Yoon Joo Youn, Kim Ji Eun

摄影师　Park Young-chae

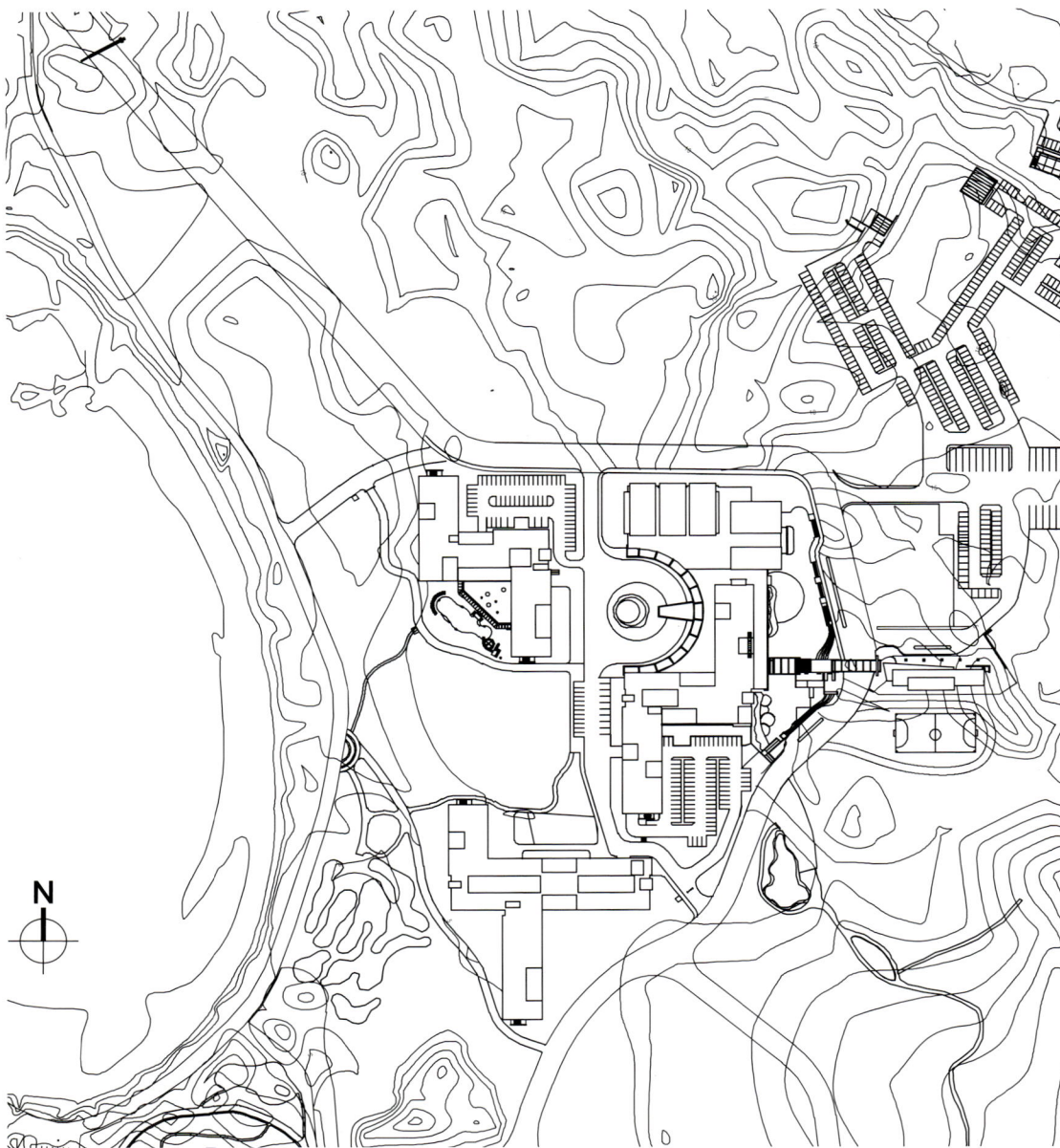

总平面图 Site plan

当一个游客沿着新铺好的黑色玄武岩大道进入济州岛凤凰度假村的时候，首先映入眼帘的建筑就是 Bella Terrace。Bella Terrace 是一套公寓，由三个住宿设施组成，红色建筑、蓝色建筑和橙色建筑。为了满足济州岛特有的家庭式旅游胜地的特点，Bella Terrace 的房间非常宽敞，有 112m² 和 179m² 两种类型。

每个房间都能看到济州岛蓝色的大海。一旦客人进入房间，建筑和房间的安排都能保证最大的独立性。Bella Terrace 的房间有华丽的内部装修，并且有各种高级的辅助设施，因此 Bella Terrace 是一个适合常住的旅游胜地。在济州岛的两个叫做 Seop-ji 和 Ko-ji 的饭店中，客人们能享受到卓越的餐厅服务。

游泳池和桑拿室视野开阔，能够看到济州岛蓝色的大海，所以客人们会感觉自己仿佛来到了赤道上的水上别墅。各种宴会场所、商店和辅助设施都让 Bella Terrace 为迎接家庭旅行者或团体旅行者做好了充分准备。

Bella Terrace is the first building to meet when a visitor would enter Phoenix Island Complex following the newly made road of black whinstone of Jeju. Bella Terrace is a condominium. It consists of three lodging facilities, Red Building, Blue Building and Orange Building. In order to meet the characteristic of Jeju Island – the staying type of family resort, the rooms of Bella Terrace are rather large, 34 pyong type(112m²) and 54 pyong type(179m²). Every room has view to the blue sea of Jeju Island. The buildings and rooms arrangement secures maximum independence, once guests would enter their rooms. Bella Terrace rooms have superb interior decoration and there are various high class auxiliary facilities so that Bella Terrace would serve as a staying type resort. Guests can enjoy excellent dining made of specialties of Jeju Island in the two restaurants, named Seop-ji and Ko-ji. Swimming pool and Sauna facility have a great open view of Jeju blue sea, so that guests would feel like they came to a certain pool villa on the equator. Various banquet facilities, stores and auxiliary facilities let Bella Terrace perfectly prepared for family travelers or group travelers.

外观 Exterior

蓝色建筑 Blue building

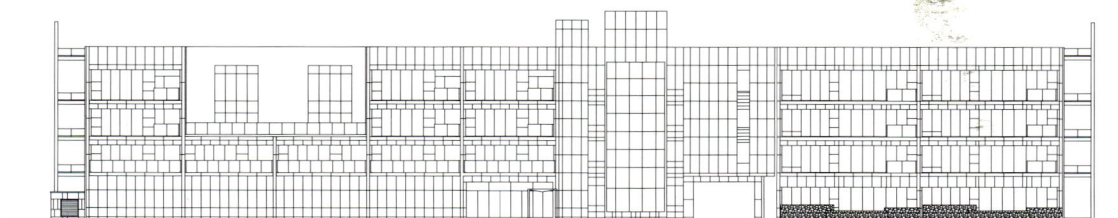

西立面 West elevation

室外与泳池 Exterior & pool

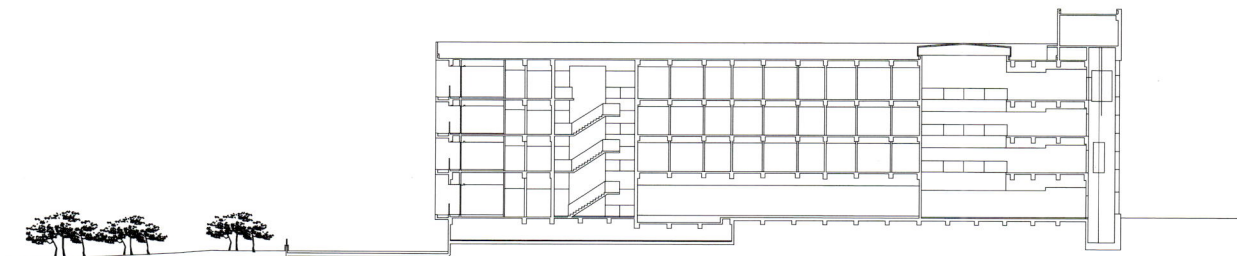

剖面图 Section

从sunken 看室外景色 Exterior view from sunken

橙色建筑 Orange building

东立面 East elevation

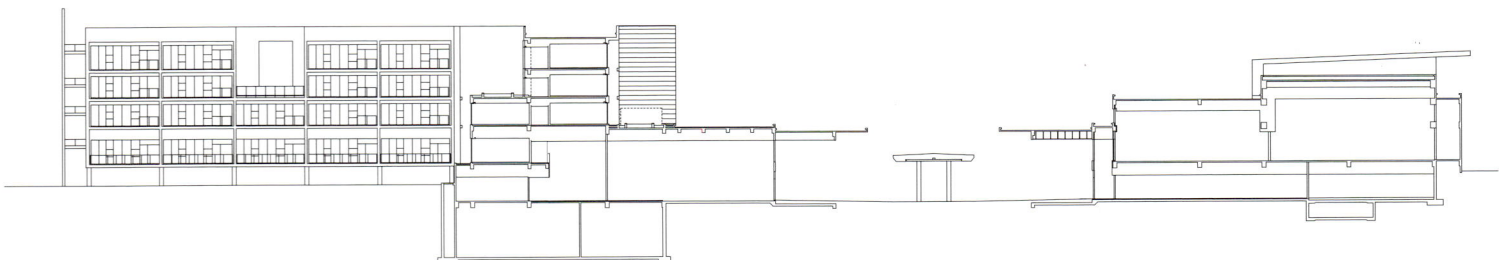

剖面图 Section

池塘 Pond

红色建筑 Red building

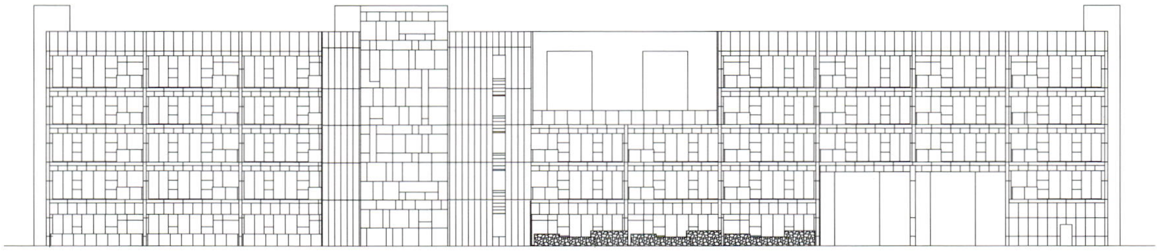

南立面 South elevation

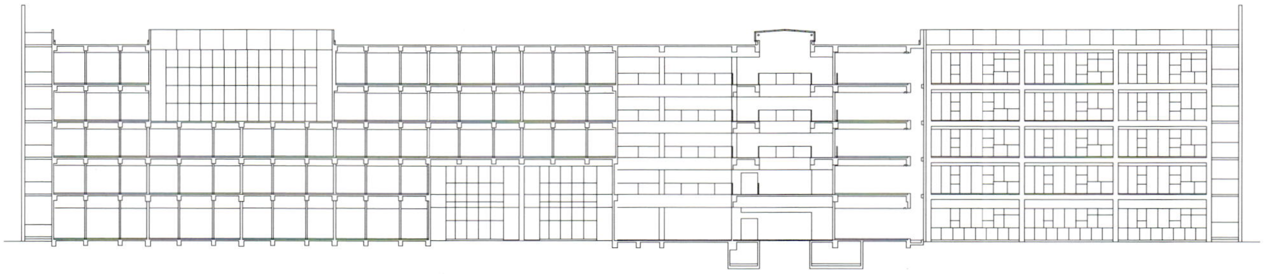

剖面图 Section

休息室 Lounge

大厅 Hall

蓝色建筑 Blue building

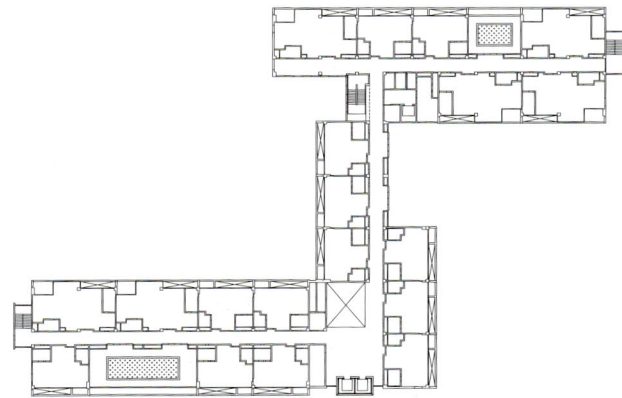

三层平面图 Third floor plan

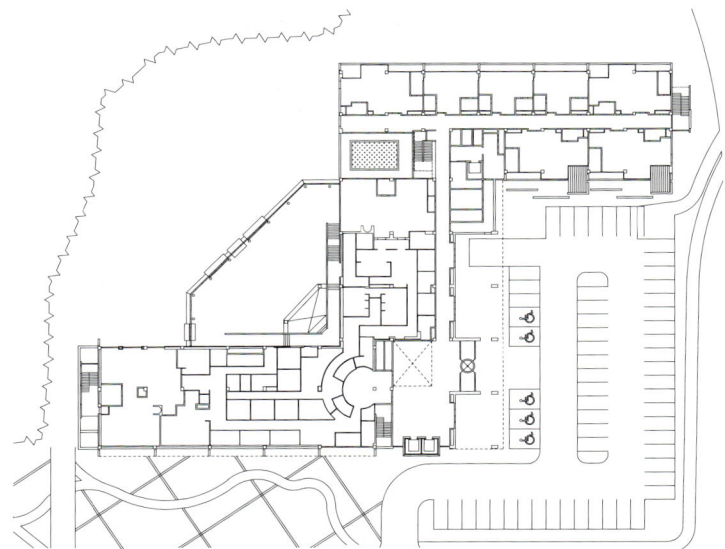

一层平面图 First floor plan

起居室 Living room

橙色建筑 Orange building

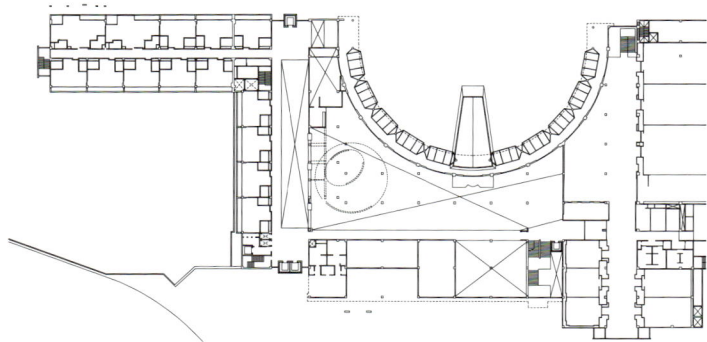

一层平面图 First floor plan

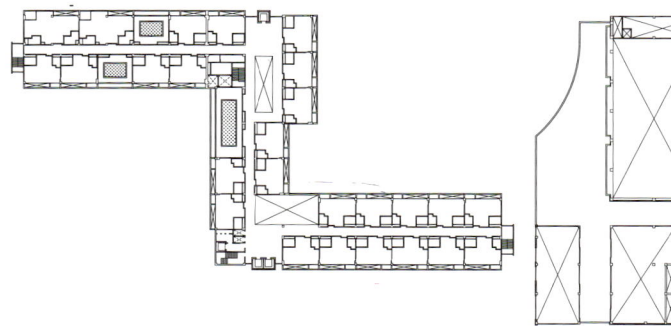

二层平面图 Second floor plan

卧室 Bedroom

红色建筑 Red building

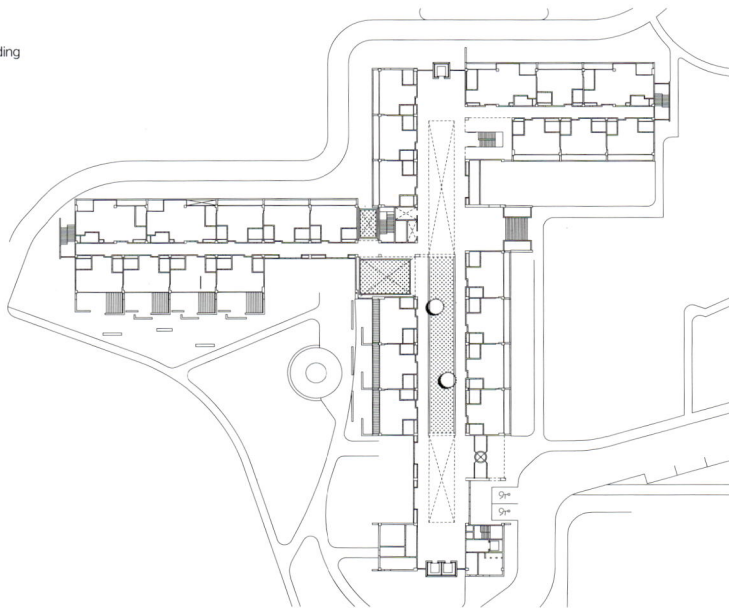

一层平面图 First floor plan

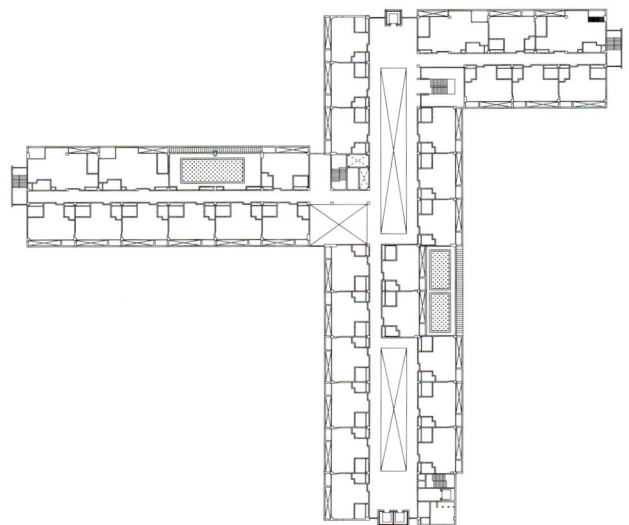

三层平面图 Third floor plan

中国威海锦湖
韩亚高尔夫俱乐部
China Weihai Point
Hotel & Golf Resort

设计 SKM Architects + Changjo Architects Inc.
地点 Weihai, China
基地面积 980,000m^2
总楼面面积 42,226m^2

威海锦湖韩亚高尔夫俱乐部位于浩瀚的海洋之滨、被联合国评为"居住环境最佳城市"的威海, 俱乐部以其独特的魅力吸引了众多人的关注。与一般的高尔夫场所不同, 以蓝色的海洋与高耸的岩石悬崖为背景的大片绿地给这个高尔夫球场营造出一种戏剧化的氛围。从划分人与自然的理念出发, 俱乐部创造了一种自然的交流方式, 展现出一个放松身心、举办聚会和商务活动的创意空间。桑拿、健身中心、餐厅等设施为客人提供了便利。

酒店位于东、西两面, 为客人提供了观看日落、日出的最佳空间。尤其值得一提的是客房, 客厅及卧室并在一起, 使空间显得更加宽敞, 浴室靠近阳台, 可以欣赏海景。功能性设计使空间效率最大化, 室内舒适的环境使客人能够愉快、舒适地放松身心。

另外, 分成 A 区、B 区、C 区的别墅富有生机与活力, 给人们带来不同的气氛。在 A 区别墅, 客人可以欣赏半岛西部边缘悬崖峭壁的全貌以及一片白色的沙滩, 气氛恬静安适, 一层与二层的四间房间都布置得非常舒适。B 区别墅呈现给客人的则是波澜壮阔的海景及宽阔平坦的道路。部分客人可置身于悬崖上方悬臂式结构的客厅内, 仿佛有种在大海中漂流的感觉。在 C 区别墅, 可以看到高尔夫球场上的活动, 视野广阔, 从而增添了视觉快感。

在风景优美的大自然中, 威海锦湖韩亚高尔夫俱乐部周围景色如画的轻松环境为现代人提供了一个体现他们紧张兴奋的生活及追求新梦想的机会。

全貌 Overall view

Located beside the endless ocean and designated by UN as "Best Residential Environment City", Weihai Point attracts the attention of many people with its own charm. Unlike general golf facilities, it conjures up a dramatic atmosphere through a broad green against the backdrop of the blue ocean and the soaring cliff of rock. The clubhouse creates a natural way of communication emerging from the concept of dividing human and nature, and presents an innovative spatial composition for relaxation, gatherings and business activities. The amenities such as a sauna, fitness center and restaurant increase the convenience of the users.

The hotel is located in the east and the west, and offers an optimized space for watching the sunset and sunrise. In the guest rooms, in particular, the living room and bedroom are merged into a single area to enlarge the space, and the bathroom is positioned near to the balcony for an ocean view. The functional design maximizes the spatial efficiency, and cozy interior offers a pleasant and comfortable relaxation.

In addition, dynamic villas classified into zones A, B, C bring about a variety of atmosphere. A one, where you can have a holistic view of west-side cliff of the peninsula and a white sand beach, brings a tranquil atmosphere and the four rooms on the 1st and 2nd floors give you a consistent comfort. B zone offers a whole view of marvelous ocean and vast fair way. Part of the mass in the living room is perched in a cantilever style over the cliff, making you feel as if you are floating in the ocean, while C zone, in which you can enjoy the play of golfers, doubles the visual pleasure with its broad view.

Weihai Point Hotel & Golf Resort provides the contemporaries with an opportunity to reflect their hectic life and dream, a new dream in a picturesque relaxing space in the beautiful landscape of nature.

外观1 Exterior I

外观2 Exterior II

度假胜地景观 Landscape with resort

外观 4 Exterior IV

从露台欣赏室外风景 Exterior & outside view from terrace

大厅 Lobby

房间1 Room I

房间2 Room II

户型平面图 1 Unit floor plan I

户型平面图 2 Unit floor plan II

1 门厅
2 走廊
3 餐厅
4 会议室
5 卫生间
6 商店
7 大厅
8 桑拿
9 前台
10 休息室
11 商务中心
12 健身中心

1 Lobby
2 Corridor
3 Restaurant
4 Conference room
5 Toilet
6 Shop
7 Hall
8 Sauna
9 Front desk
10 Lounge
11 Business center
12 Fitness center

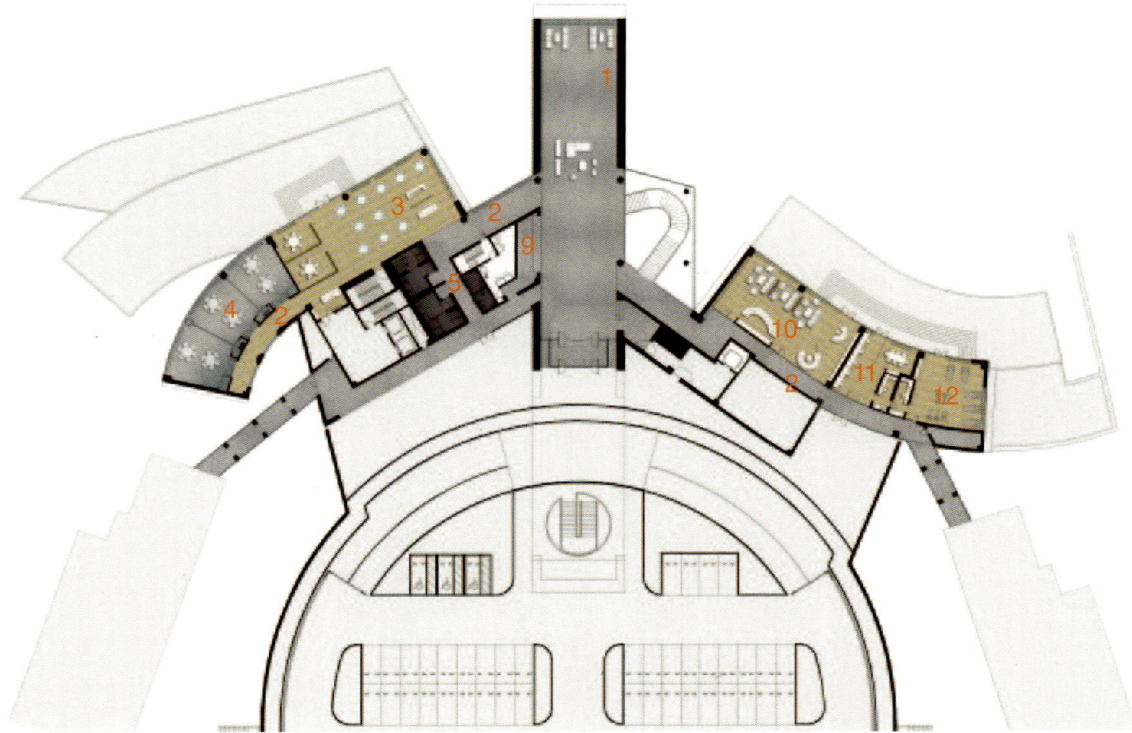

二层平面图 Second floor plan

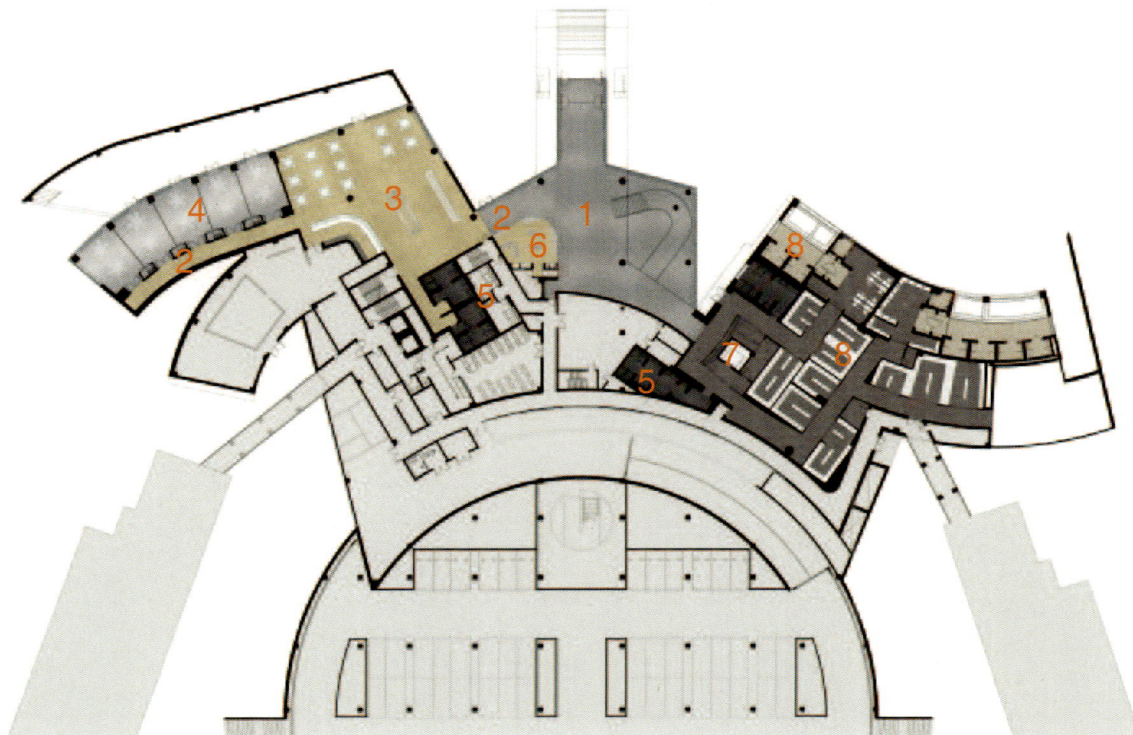

一层平面图 First floor plan

KUUM 温泉住宅酒店
The KUUM Hotel Spa & Residences

地点 GAD / Gokhan Avcıoglu
Bodrum, Turkey

基地面积 20,000m^2

总楼面面积 20,000m^2

甲方 BIRTUR / Tuncel Group

室内设计 T Architecture(hotel room, restaurant, meeting room, spa hammam)

景观设计 Zeynep Akgoze

监理 Ali Terzi

摄影师 Ali Bekman, Ozlem Avcıoglu, Leo Fabrizio

KUUM 温泉住宅酒店是精品生活及度假的创新理念。KUUM 的设计灵感来源于当地的美景、资源及土耳其西南部及爱琴海的历史遗产。尤其值得注意的是，KUUM 酒店位于国际性旅游胜地博德鲁姆，这里以其怡人的气候、碧波荡漾的海水及世界性文化氛围而著称。

博德鲁姆是地中海的贸易港口，位于土耳其西南部，该地区有三千多年的悠久历史，其中包括希腊时代。这里是备受尊崇的科学家 Heredot 的诞生地，许多艺术家如 Leochares、Bryaxis 及 Timotheos 的雕塑作品都曾在这里展览，如今收藏在世界各地的博物馆里。

设计师将 KUUM 建成一个依靠海湾的小型建筑群，和山坡地势相辅相成。甲方希望这个酒店在设计和建造方面，与博德鲁姆的普通度假区相比能够别出心裁，而严格的建筑条例也要求设计需要创新、灵活。

"The KUUM Hotel Spa & Residences" is an innovative concept for boutique lifestyle and resorts. The architecture for "KUUM" was inspired by the regional beauty, resources and historical legacy of southwestern Turkey and the Aegean Sea. Specifically, the Kuum Property is located in Bodrum, an international destination renowned for its wonderful climate, turquoise waters and cosmopolitan atmosphere.

Bodrum is a Mediterranean port-trade settlement in the Southwest of Turkey. The area boasts a rich history of over three thousand years, including Hellenistic times. The venerated scientist Heredot was born there and sculptures by artists including Leochares, Bryaxis and Timotheos were exhibited there and can now be found in museum collections around the world.

"Kuum" is conceived as a small settlement, nestled on a seafront bay, the property uniquely coexists within the hillside topography. The clients wanted a design that looked and performed differently than what is normally expected of resorts for Bodrum. Strict building codes also demanded innovative and flexible approaches to design.

示意图 Diagram

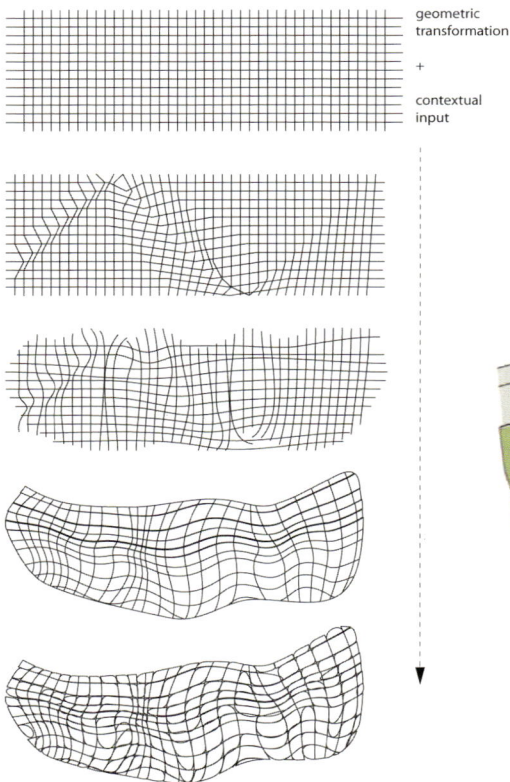

geometric
transformation

+

contextual
input

总平面图 Site plan

设计进展 Design evolution

terrain

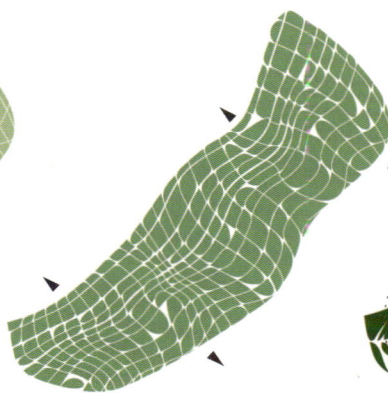

terrain sculpting - areas of deformation

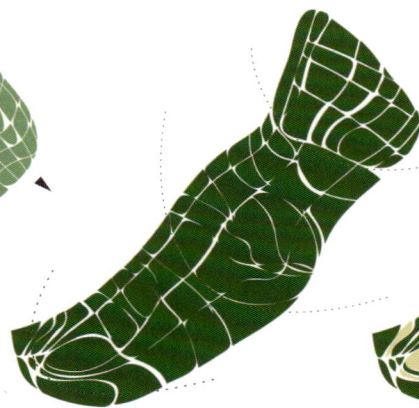

terrain scalloping

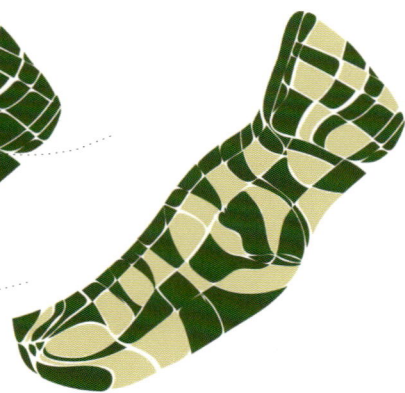

terrain sculpting - siting, programming

distinguishing programs

softscape - vegetation

built/hard scape

layering

甲板酒吧（夜景）1 Deck bar(night) I

甲板酒吧（夜景）2 Deck bar(night) II

酒店建筑的形变 Hotel block deformation

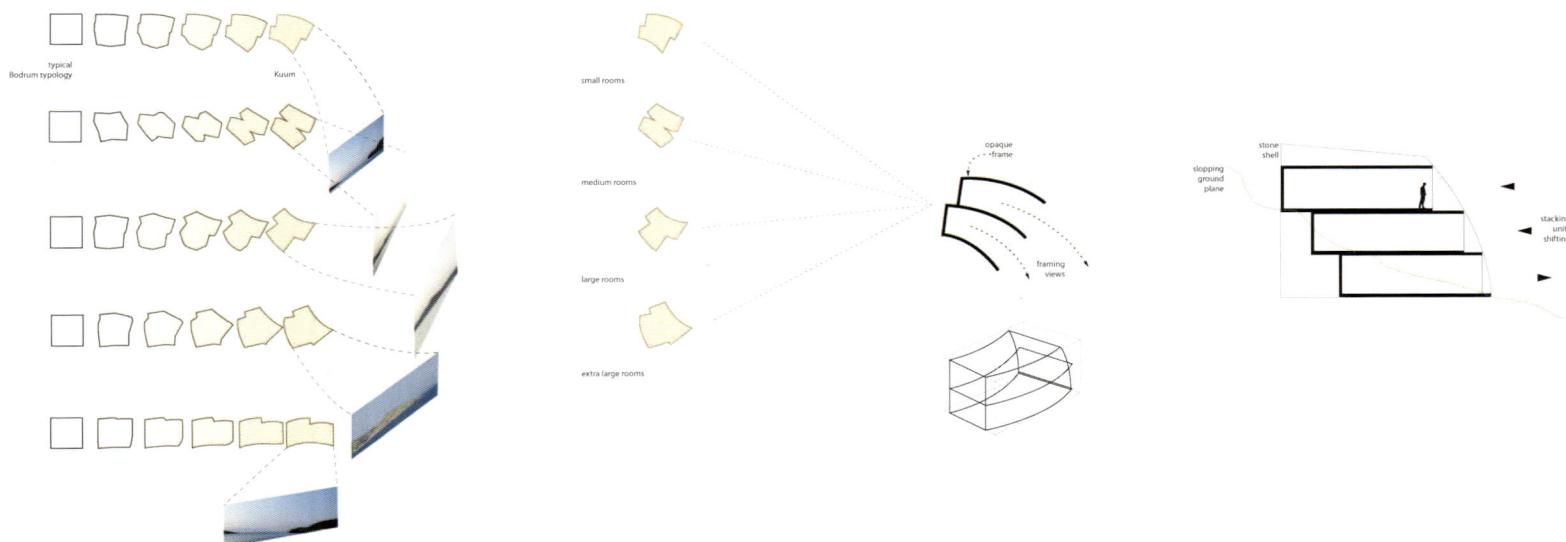

typical
Bodrum typology   Kuum

small rooms

medium rooms

large rooms

extra large rooms

opaque
+frame

framing
views

stone
shell

slopping
ground
plane

stacking
units
shifting

甲板酒吧（夜景）3 Deck bar(night) III

surface techinques for hotel & residence buildings
stone variations

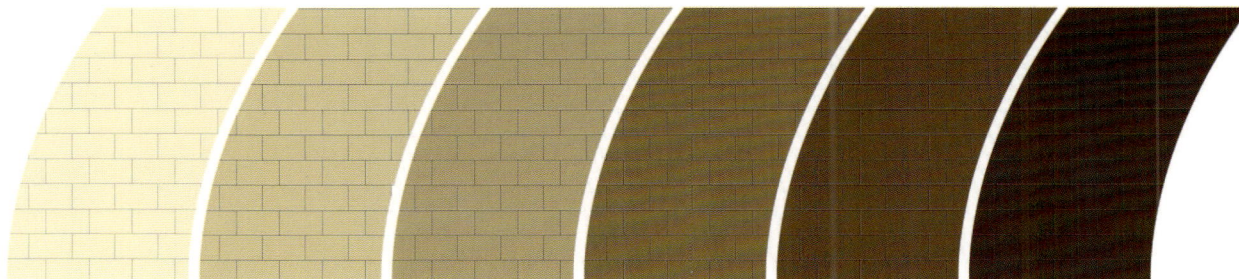

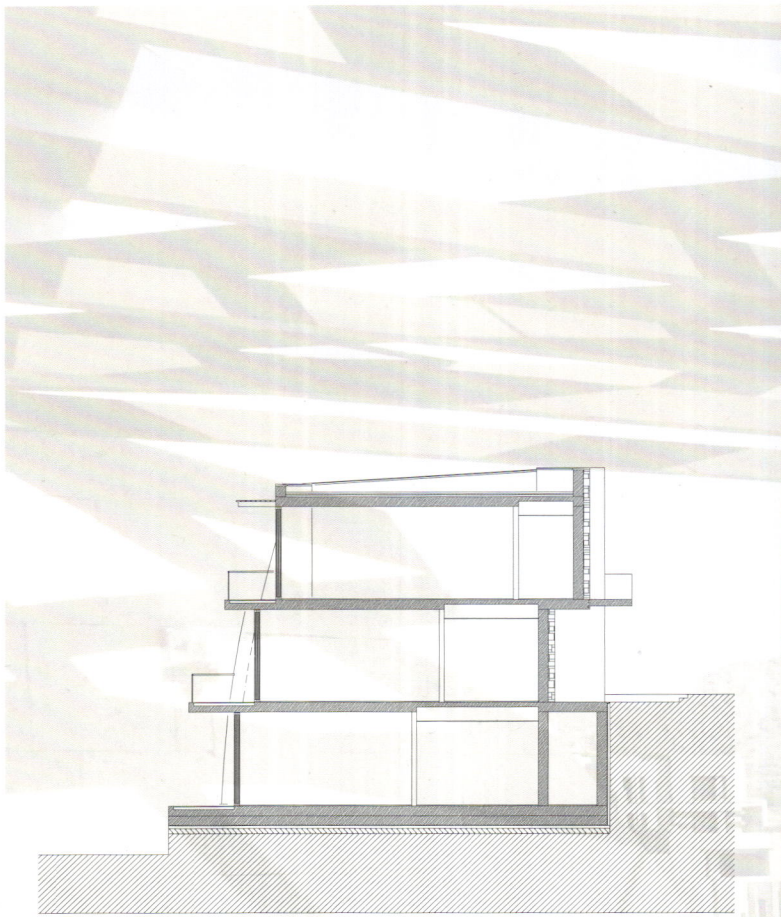

剖面图（酒店）1 Section(Hotel) I

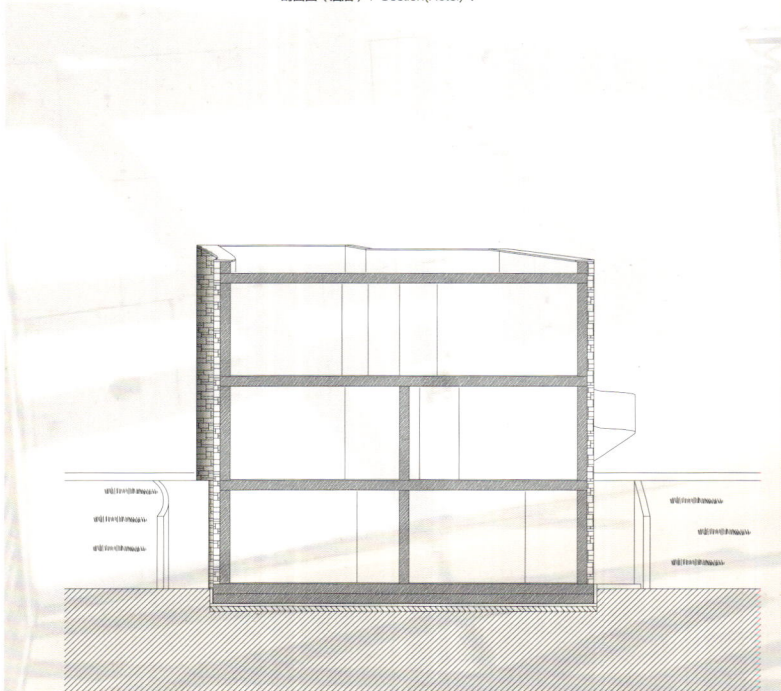

剖面图（酒店）2 Section(Hotel) II

甲板吧 Deck Bar

外观 1 Exterior 1

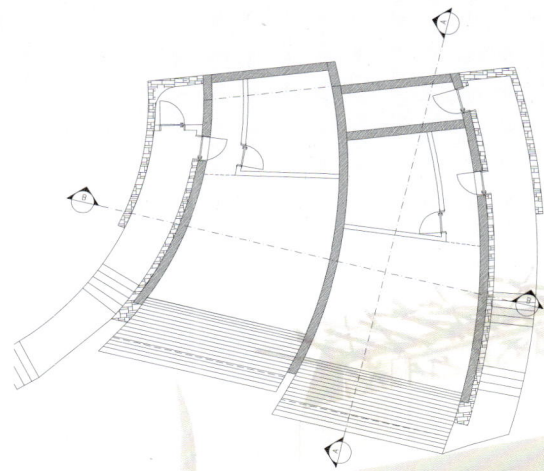

平面图（酒店）1 Plan(Hotel) I

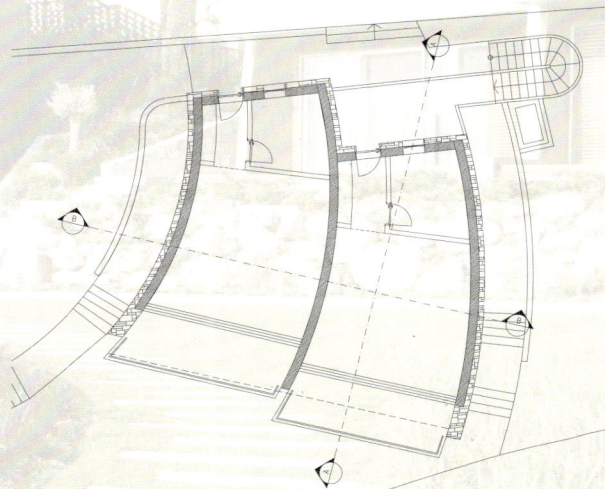

平面图（酒店）2 Plan(Hotel) II

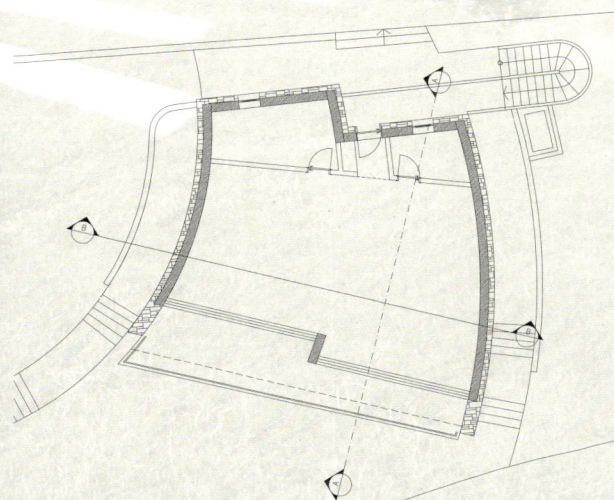

平面图（酒店）3 Plan(Hotel) III

从室内欣赏室外风景 View from inside to outside

平面图（海滨餐厅） Plan(Pier-restaurant)

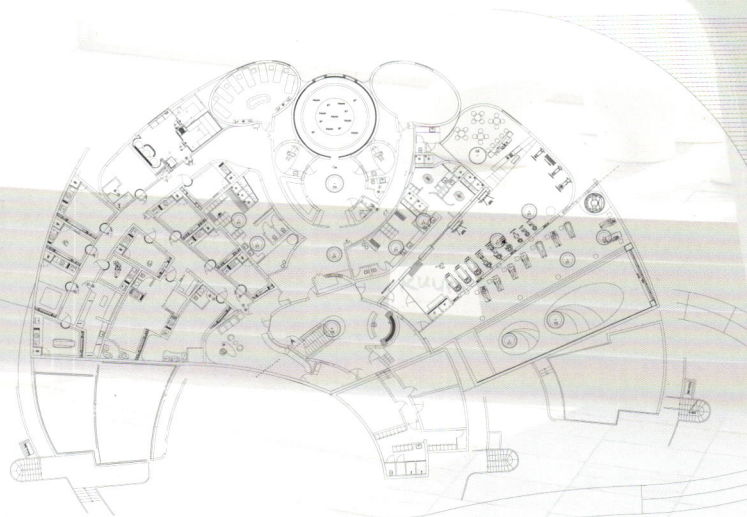

平面图（温泉浴场） Plan(Spa)

Amangiri 度假村
Amangiri

设计　Rick Joy + Marwan AL-Sayed + Wedell Burnett
地点　Southern Utah, America
摄影师　offered by Amangiri

在深度了解西南大沙漠的基础上，Amangiri 的设计理念是创造一个能够真切体验沙漠环境的场所，不借用单纯的文化意义，而是注重这个特殊地点和场所更加真实的一面，即包围 Amangiri 度假村的景观及光线。

度假村位于一片砂岩层低处，就像一个古代的建筑群，内设 34 间房间。客人在每间房间都能欣赏到度假村周围平顶山及当地整日沐浴着阳光的那种未经雕琢、天然纯粹的优美风景。在度假村清晰的轮廓之外，只剩下度假村所在的那片原始的地势。度假村最受关注的地方是内设起居室、餐厅、图书室及画廊的楼阁，及环绕天然岩石绝壁的壮观游泳池。在这里同时强调了池水、岩石、天空这三项构成这个景观的基本元素。这些建筑以有棱角的简约风格设计而成，简易的混凝土砌体按照功能、移动和光线进行修建：被冻结的永恒的体量结合在一起，渲染出一个抽象的地形，颜色很轻易地与沙子、鼠尾草、岩石构成的不断变化的景观融合在一起。从楼阁及主游泳池可分别通向两个背依岩石而弯曲的建筑侧翼。楼阁东侧的沙漠之翼由 16 间套房构成，可通过一条有外墙的小巷到达。小巷设计得如同峡谷一般，置身于此，耳中充溢着水流的自然之声，感受翠绿苔藓散发出来的潮湿气息。南面的平顶山之翼有 18 间套房，在沙漠与起伏的岩层中伸展开来。

经过岩石形成的拱门及有遮掩的私人庭院可以进入套房，石拱门令人联想起当地峡谷的奇观。套房从庭院开始逐渐展露出来，给客人提供了观赏周边沙漠壮观景色的视野。每间套房中央凸起的石头岛屿上都设有一张床、桌子及沙发。屋外是沙漠中私人的休息区，在私人小火炉周围另设躺椅，奢华的环境给人一种露营的感觉。单独的浴室及更衣室是由将房间分隔开的密实石块雕刻而成的，绿色的石头、经过过滤的灯光和池水与强光及起居室和卧室内的奇景形成对比，从而达到了一种神秘的效果。个别套房带有私人泳池以及天台，天台上有休息用的床，白天可以躺在上面休息，晚上又可以看星星。建筑两个侧翼的另一端是更加宽敞的 Amangiri 套房及 Girijaala 套房，客厅、卧室、餐厅、卫生间都非常宽敞，还有宽大的私人泳池及天台。

Aman 温泉会馆位于平顶山一翼内，与数千年来由大自然塑造的岩层奇观亲密接触。温泉包含五个单独的楼阁及水池，内外的设计目的都是为了给客人提供私下沉思的环境。温泉反射出周围岩层永恒的本质，岩层倒映在水中，仿佛抽象的滚动的岩石。根据功能和位置的不同，岩层的倒影也随之变化，或重或轻，或呈固态，或呈液态。淋浴区雕刻成有机的形式，饰以神秘而自然的灯光或彩灯。烘干区内部为木质结构，灯光宁静安详。

Based on a deep understanding of the Desert Southwest, the brief was to create a place of authentic experience, not only based on simplistic cultural appropriations, but rather on what was most true to this particular site and place – namely the landscape and the light that envelops it.

The 34-room resort is situated against a low entrada sandstone rock formation, rather like an ancient settlement. From every room guests can appreciate the rawness and pure natural beauty of the surrounding mesas and the region's mesmerizing light play throughout the day. Beyond the clearly defined line of the resort there is nothing but the pristine terrain within which Amangiri is set.

The focus of the resort is the Pavilion which houses the Living Room, Dining Room, Library and Gallery, and the spectacular swimming pool, which wraps around a natural rock escarpment. Here the basic elements of this landscape are juxtaposed and emphasized: water, rock and sky. The buildings are designed with angular minimalism, simple concrete blocks carved by program, movement and light: frozen, timeless mass is rendered as abstract geology, with colours that blend effortlessly into the shifting landscape of sand, sage and rock. Leading from the Pavilion and main swimming pool are two separate wings that bend and fold against the rock. The Desert Wing to the east of the Pavilion is composed of 16 suites that are reached via an external walled lane. Designed as an abstraction of a slot canyon, the lane is replete with the natural sound of water and the moisture of verdant moss. The Mesa Wing to the south features 18 suites, unfolding across the desert sand and undulating rock formations.

Suites are entered via rock archways, reminiscent of the region's spectacular slot canyons, and private screened courtyards. From the courtyards, the suites unfold, affording guests breathtaking framed views of the surrounding desert scenery. A raised stone island in the centre of each suite incorporates a bed, desk and sofa. Outside, a private desert lounge provides additional lounging benches around a private fire pit, capturing a sense of camping under the stars in a luxurious setting. The separate bathing and dressing areas are carved out of the dense stone mass that separates individual rooms. The effect is one of mystery, with green stone, filtered light and water elements contrasting with the bright light and striking views of the living and sleeping areas. Several suites are designed with private pools, as well as sky terraces featuring lounge beds for relaxing by day, or star-gazing by night. At the opposite ends of both wings are the larger Amangiri and Girijaala Suites. These offer spacious living, sleeping, dining and bathing areas, as well as generously-proportioned lap pools and extensive sky terraces.

The Aman Spa, located within the Mesa Wing, engages directly with the wonders of rock formations crafted over millennia by wind and water. Incorporating five separate pavilions and water elements, the Aman Spa is designed for intimate reflection, both inside and out. The architecture mirrors the timeless nature of the surrounding rock formations: the pavilions scattered like tumbled rocks, abstracted, and made solid or liquid, heavy or light depending on program and placement. Wet treatment areas are defined by sculpted organic form and mysterious, natural or coloured light, while dry treatment areas are defined by wood linings and serene light.

全景 Overall view

外观 1 Exterior I

游泳池 1 Swimming pool I

游泳池 2 Swimming pool II

沙漠中的套房休息区 Suite desert lounge

黄昏时的沙屋休息区（夜景）Desert lounge dusk(night)

休息区（夜景）Studio lounge(night)

沙漠中的套房休息区（夜景）Suite desert lounge(night)

餐厅 Dining

起居室 Living room

浴室风光 View from bath shower

瑜伽健身室 Yoga pavilion

Aman 温泉疗养室 Aman

客房 Guest room

卫生间 Bathroom

江苏金诚酒店与度假村
Jiangsu Gold Sense Hotel & Resort

设计	Samoo Architects & Engineers / Sohn Myung-gi, Hwang Han-gu
地点	Jiangsu, China
用途	Sightseeing & Rest institution
基地面积	22,197m^2
建筑面积	12,700m^2
总楼面积	148,300m^2
楼层	B4, 48FL
结构	Reinforced concrete, Steel, Steel truss
饰面材料	Steel panel, Glass curtain wall(ext.), Stoneware, Carpet(int.)

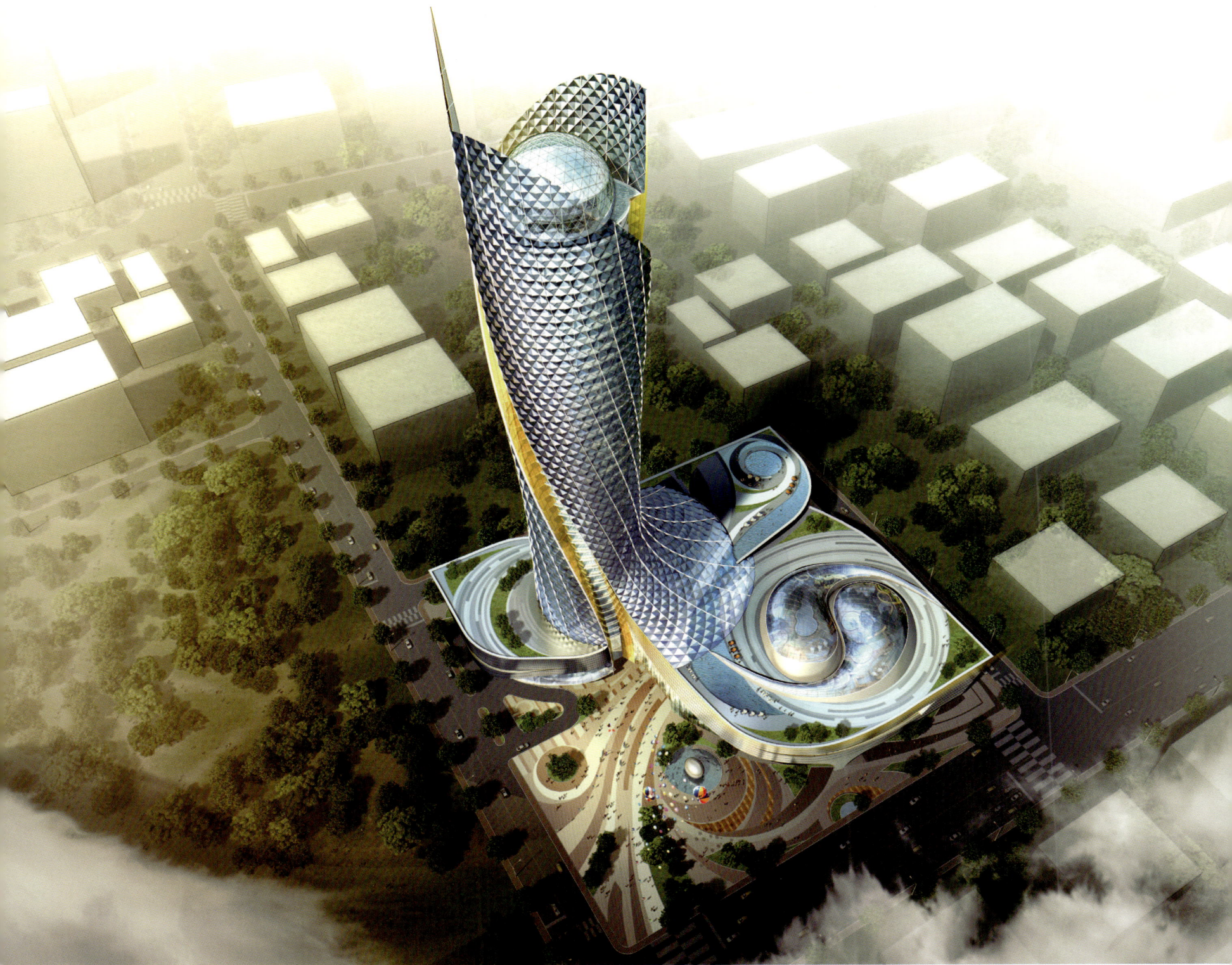

龙井柱园

江苏省江阴市是中国核心城市之一，同时也是扬子江经济区的重要物流配送中心。由于中国中央政府及江阴市政府对该地区的发展给予了强大的支持，所以许多国外企业及国内公司都在江阴市设立了办事处。目前，江阴市对文化生活及消费支出方面需求的期望值很高。

项目业主金诚国际控股集团有限公司要求酒店的设计要独特、有创意，使其成为全球性的标志性建筑。业主要求建设都市度假村形式的设施，使市民及游客可以聚在一起相互交流。在设计方案中，除了已经被他人规划的设施外，我们还考虑到扬子江、运河及城市绿化带等周边的自然环境。最终提案成功地将业主要求的都市度假村的理念与韩式娱乐设施的理念一同囊括在内。

我们的方案建议将酒店分为五个区来体现设计的理念，达到有效的空间规划，并使客人活动路径的规划达到最佳。这五个区包括酒店、游乐场、水上公园、娱乐区及综合设施区。而设计的最重要目标是其独特性及创新性，使酒店作为江阴市的一个标志性建筑，成为一个典型的都市度假村。

设计过程中，我们面临的最大挑战是如何创建一个能够代表中国特色的典型符号，而建筑内部则需体现出韩国的文化及韩流理念。建筑的上部刻画了北京天坛公园的龙井柱（中国传统的柱形设计）图形，另一方面，位于建筑下面部分的水上公园的屋顶为太极形状，象征着韩流风尚，这两个设计元素以其流畅的外形融入整个建筑中。夜晚，代表中国的两种典型颜色——金色与红色散发出的光芒可用来照明。玻璃幕墙立面上安装的照明灯能够在夜晚呈现各种生动的夜景。玻璃幕墙上为菱形图案，中国另一种典型形象——龙的皮肤。酒店共48层，屋顶设有大球形的露天泳池，从这里能够看到江阴市的全貌，这个泳池将成为整个建筑的标志。计划将五星级酒店设在建筑的上面部分，而下面部分将设水上公园、娱乐区（带有剧院、电影院）、综合设施区（带有大型购物中心）及游乐场。这些娱乐设施将通过单独的分区规划有机地联系起来。

Long-jing-zhu-yuan(dragon-pond-column garden)

Jiangyin City of Jiangsu Province in China is one of core cities and an important distribution hub of Yangzhi River economy zone. As Jiangyin city government and Chinese central government give strong support on the development in this area, many foreign and Chinese companies are establishing offices in Jiangyin City. At present, there is a high level of expectation on the demands in culture life and consumer expenditure in Jiangyin City.

The project owner Gold Sense International Holdings requested the design to be highly creative and unique enough to be a global icon. The owner required an urban-resort type facility where citizens and travelers can get together and socialize with each other. In the design proposal, we considered the surrounding natural environment such as Yangzhi River, canal and green belt in addition to other facilities already being planned by others. The final submitted proposal successfully contained the concept of an urban-resort concept required by the owner together with Korean style entertainment facilities.

Our proposal suggested five zones to reflect the design concept and achieve efficient space planning and optimized movement-line planning. The five zones are: Hotel, Casino, Water Park, Entertainment Zone and Multi-complex. Still the most important objective was a unique and creative design so that the hotel would become a representative urban-resort as a landmark of Jiangyin City.

Our biggest challenge during the design was how to establish the strong symbolism representing China, while expressing Korean culture and Korean Wave concept in the architecture. The upper part of the building visualized the Long-jing-zhu(dragon pond column) Tian-dan Park in Beijing, which is traditional Chinese column design. On the other hand, the roof of the Water Park in the lower part of the building visualized the shape of Taegeuk, which symbolizes the Korean Wave. These two elements are integrated in the building with smooth and streamlined shape. The two representative colors of China, gold and red colors, are used for the nighttime lighting. The LED lighting installed on the facade of the curtain wall will give various and colorful night views in the night. The curtain wall has lozenge pattern, which brings up the image of dragon skin, another representative image of China. On the rooftop of the hotel with 48 floors, an Outdoor Swimming Pool with a shape of a big ball is planned. This Swimming Pool will be able to have a great panoramic view of the city and will become the symbol of the building. The floor planning located a five-star hotel in the upper part of the building. In the lower part, there will be: Water Park, Entertainment Zone (with theater and cinema), the Multi-complex(with large scale shopping mall) and the Casino. These entertainment facilities will be connected organically by the separate Zoning Plan.

立面图 Elevation

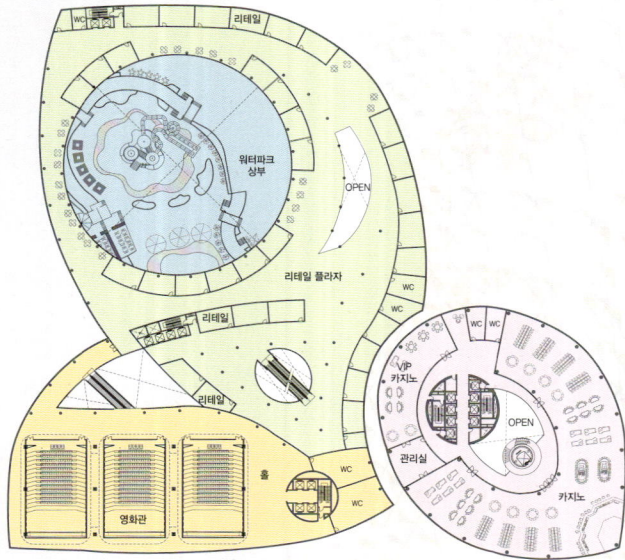

三层平面图 Third floor plan

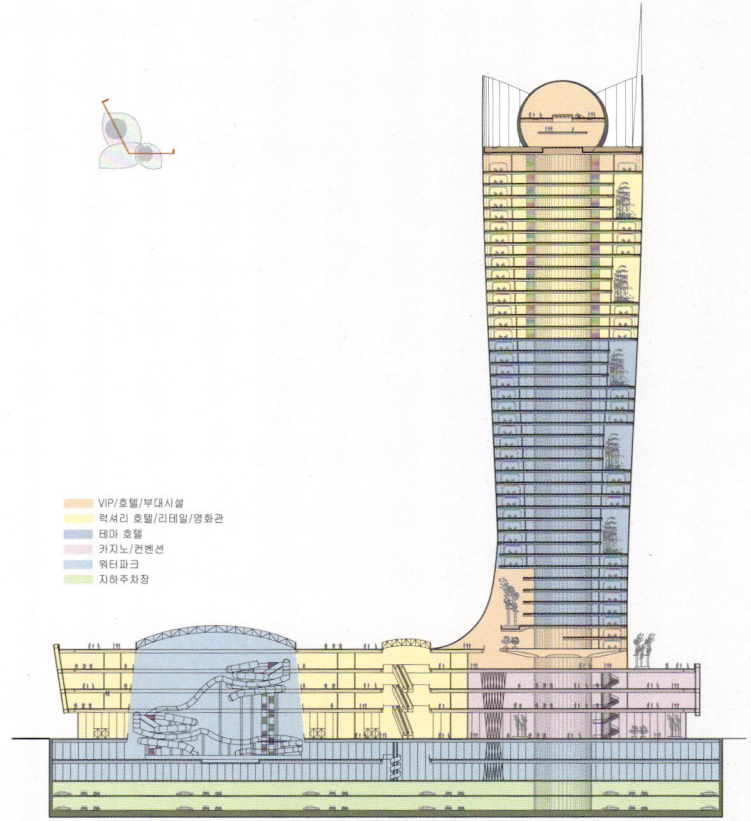

리테일
WC
워터파크 상부
OPEN
리테일 플라자
리태일
WC
VIP
카지노
OPEN
관리실
WC
WC
카지노
WC
홀
WC
영화관

剖面图 Section

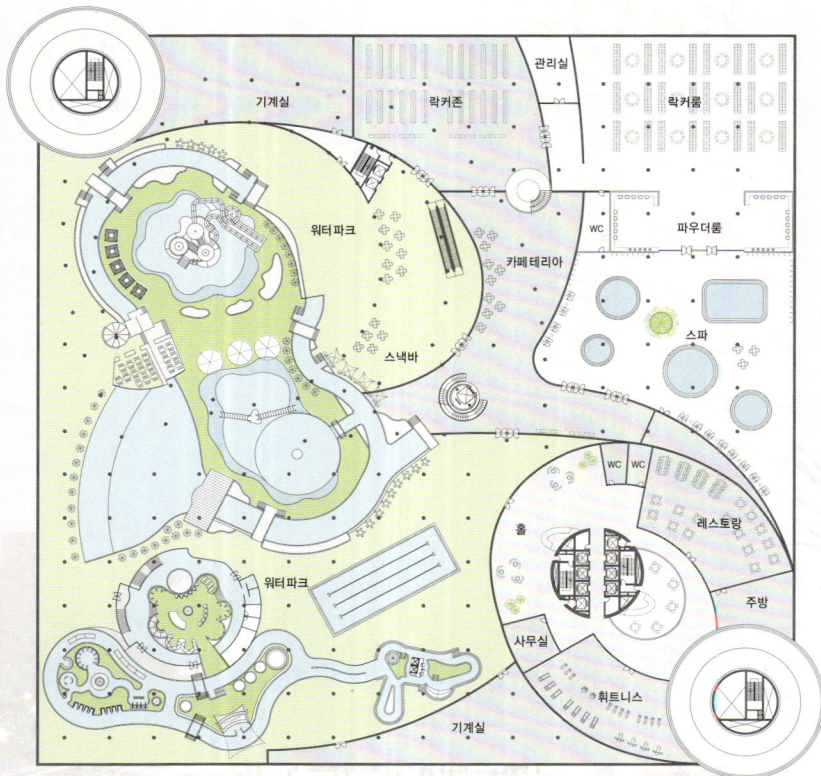

VIP/호텔/부대시설
럭셔리 호텔/리태일/영화관
테마 호텔
카지노/컨벤션
워터파크
지하주차장

地下一层平面图 Basement first floor plan

기계실
락커존
관리실
학커룸
워터파크
카페테리아
WC
파우더룸
스낵바
스파
워터파크
홀
레스토랑
WC WC
주방
워터파크
사무실
휘트니스
기계실

一层平면图 First floor plan

30M 주도로
WC
리태일
리태일
보행자 주진입
워터파크 상부
12M 부도로
로비
18M 부도로
산봄언공
차량 출입구
워터파크 주출입구
보행자 부진입
차량 입구
호텔/카지노 부출입구
호텔/카지노 주출입구
리태일
엔터테인먼트 주출입구
호텔 로비
차량 출구
부드코트
지원시설
12M 부도로

MOW 俱乐部
CLUB MOW Club House

设计　Jeon-In CM Architects Associates / Ahn Myuung-jei
地点　Mogok-ri Seo Myeon Hongcheon
基地面积　11,685m²
建筑面积　3,990m²
总楼面面积　8,502m²
楼层　B1, 2FL
参与设计人员　Hwang Gyu-ha, Kang Dae-jun,
Kim Jin-woo, Han Jung-hoon, Kim Yeon-ok

总平面图 Site plan

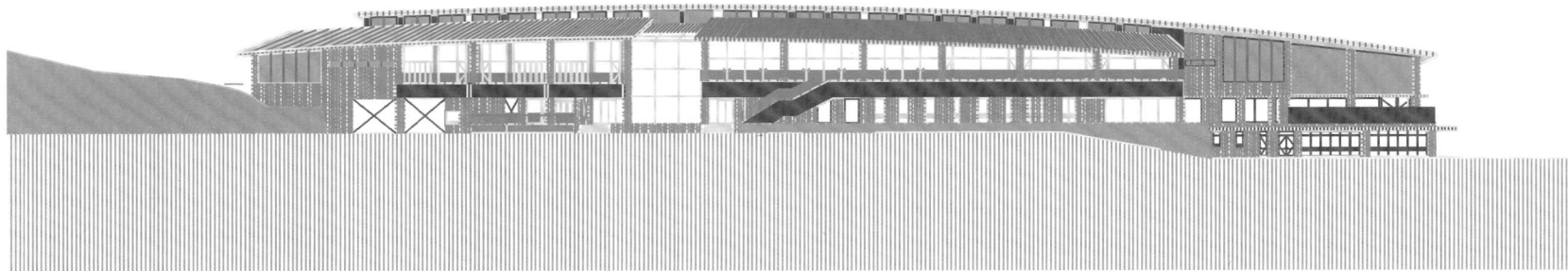

后立面 Rear elevation

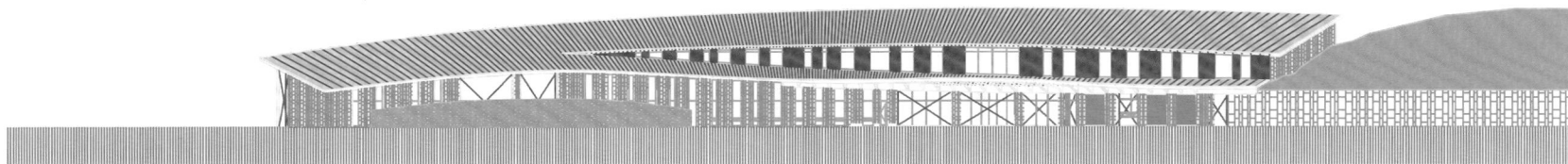

前立面 Front elevation

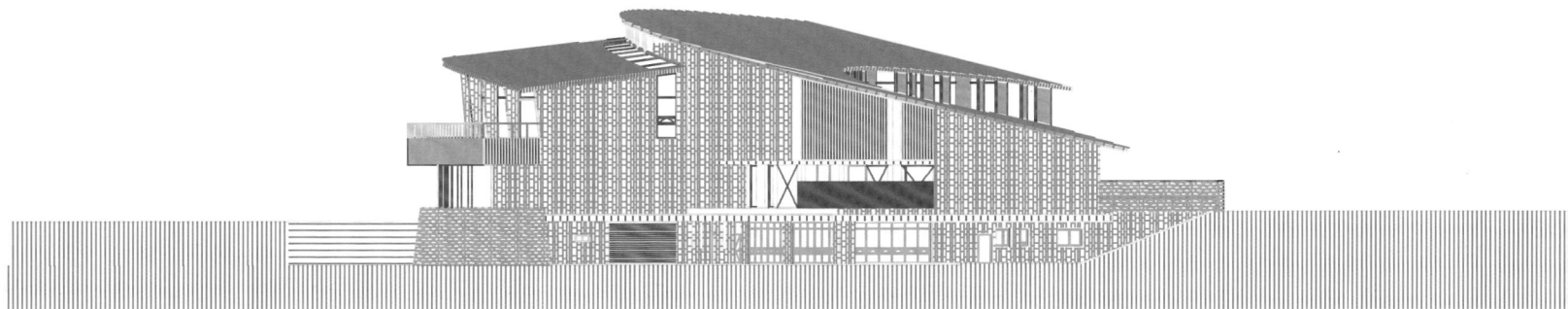

左立面 Left elevation

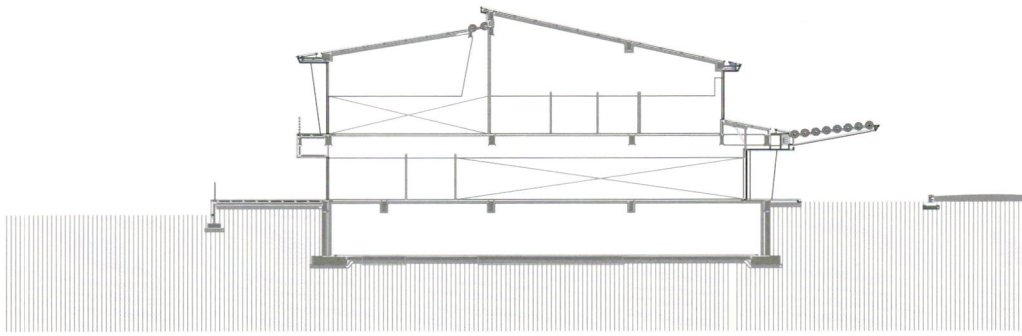

纵剖面 Longitudinal section

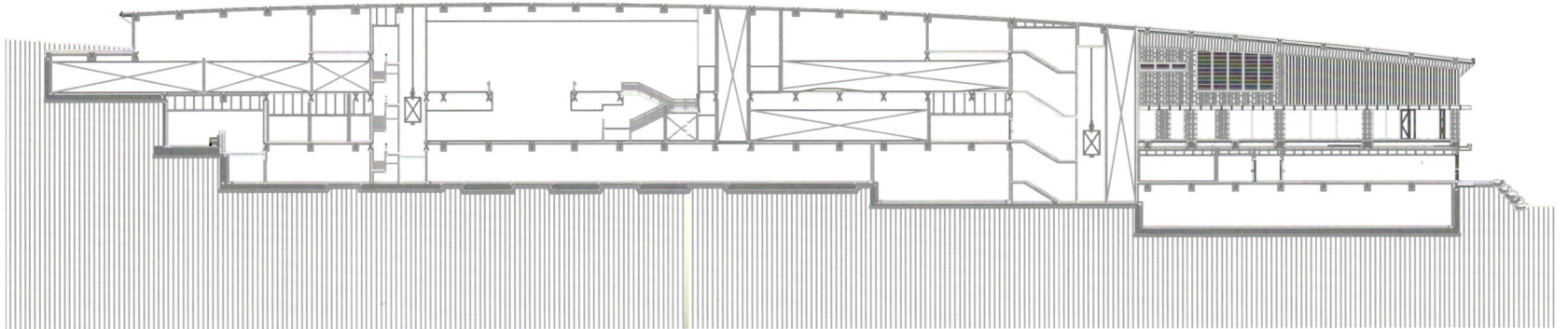

横剖面 Cross section

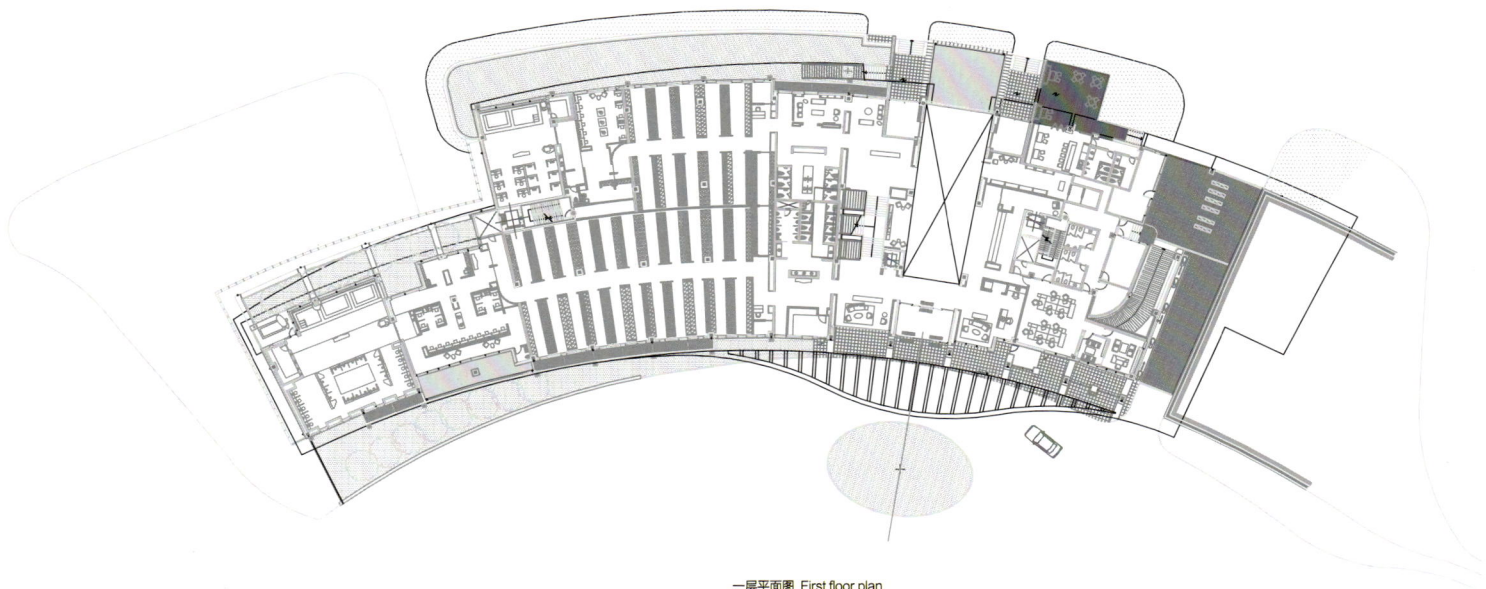

一层平面图 First floor plan

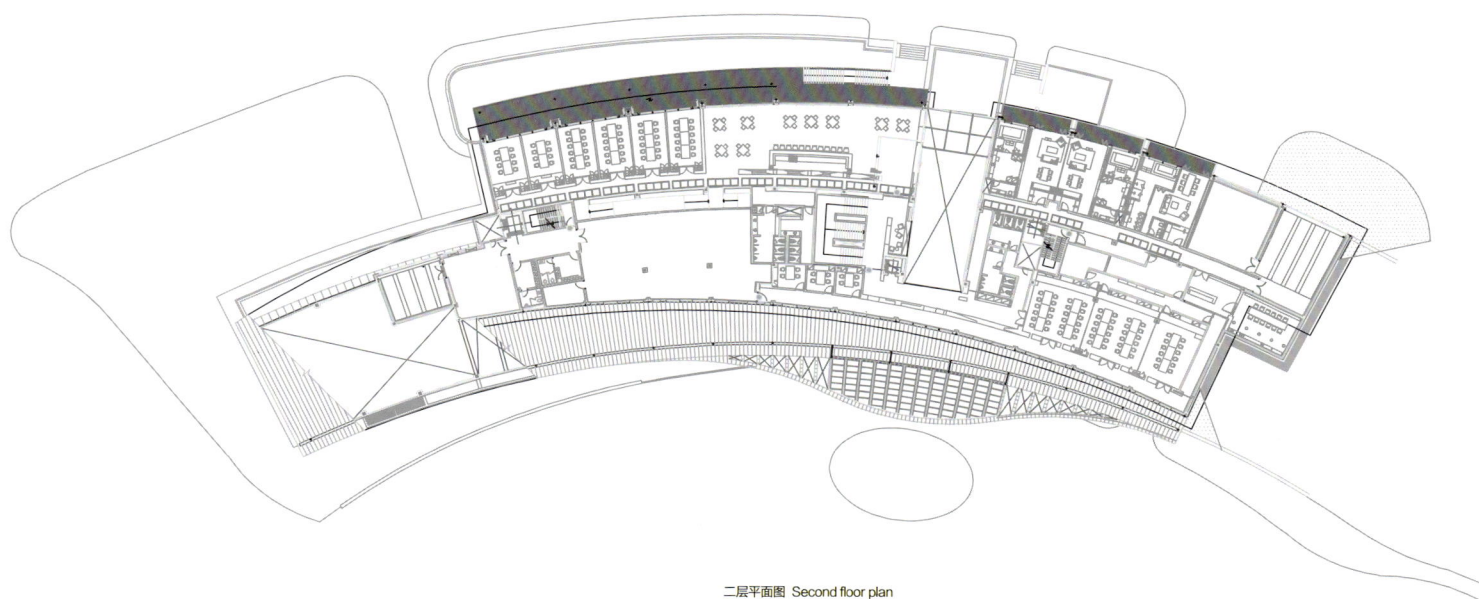

二层平面图 Second floor plan

Gele 海滩与度假村
Gele Beach and Resort

设计 FREEFORM+DEFORM
地点 Alpha Beach, Nigeria
基地面积 44,515m^2
甲方 Almat Land services Ltd.

总平面图 Site plan

立面图 Elevation

Gele 海滩与度假村是一个 44 515m² 的城市开发与滨海方案的项目，其设计目的是以尼日利亚丰富的文化氛围及优美而开发不充分的海滨吸引各方在其潜能巨大的旅游业方面投资。

这个开发项目在外形与象征意义上将滨海活动与都市构造联系起来，该都市构造与高端零售、娱乐及提供社区、文化教育及休闲机会的餐饮环境结合在一起。项目的总体规划及作为项目零售部分的大型起伏状格子顶棚通过图案的替换与变化勾勒出项目的方案设计及外形设计图，这种替换与变化与非洲女子戴在头上的传统头巾的折叠方式相像，项目便因此而得名。

Gele 海滩建筑能够针对各种活动提供大量丰富的体验空间。从风景优美的天然园林、树木成排的散步广场、空气清新的公园和水景特色到高端零售店、充满生机的露天市场、充溢着艺术、文化、技术的热闹夜生活，Gele 海滩力争将所有灵感聚集在一处，吸引游客及当地居民前来体验。

"Gele Beach and Resort" is a 44,515 m² urban development and beach-front proposal designed to attract investment in the abundant tourism potential of Nigeria's rich cultural offerings and beautiful yet underdeveloped ocean beaches.

The development is concerned with physically and symbolically connecting the activities of the ocean front to the urban fabric which combines hi-end retail, entertainment and dining environments with the socially programmatic ambitions of providing community, cultural education and leisure opportunities. The master plan and specifically the large undulating trellis canopy which anchors the retail portion of the project, draws its conceptual and formal design from the way the patterns shift and transform with each fold of fabric in the traditional headdress worn by African women for which the project is named.

The architecture of "Gele Beach" was developed to provide an abundance of rich and experiential spaces for a variety of activities to occur. From beautiful natural gardens, tree-lined promenades and refreshing park and water features to hi-end retail outlets, vibrant open-air markets and a thriving night-life brimming with art, culture and technology, "Gele Beach" seeks to aggregate all that inspires to engage the human spirit for both tourists and local residents alike.

高架线 High line

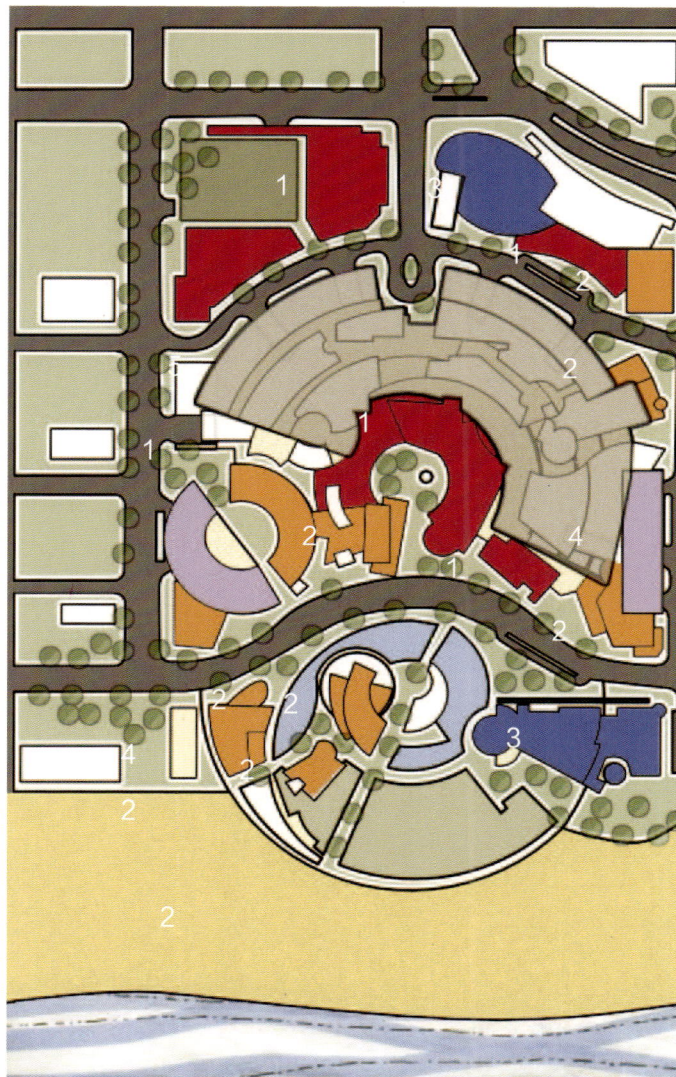

1 零售店
2 餐厅
3 中央区域 / 剧院
4 酒店 / 分时度假区
5 停车场

1. Retail
2. Restaurant
3. Center / Theatre
4. Hotel / Timeshare
5. Parking

一层 Level 1

1 零售店

2 餐厅

3 中央区域 / 剧院

4 酒店 / 分时度假区

5 停车场

1. Retail

2. Restaurant

3. Center / Theatre

4. Hotel / Timeshare

5. Parking

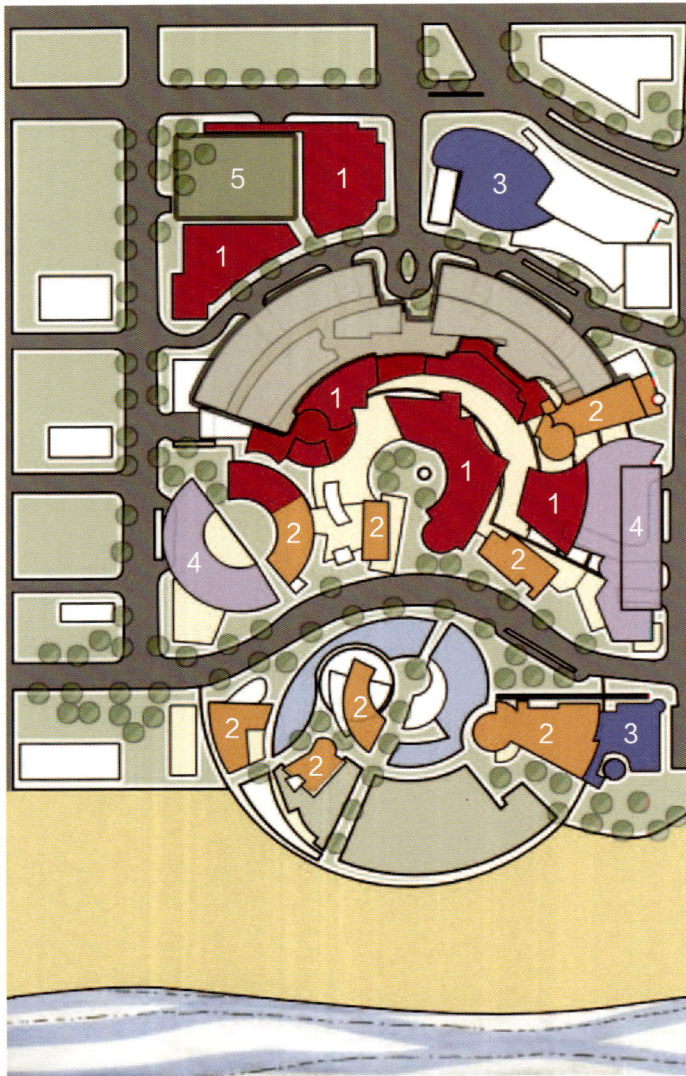

二层 Level 2

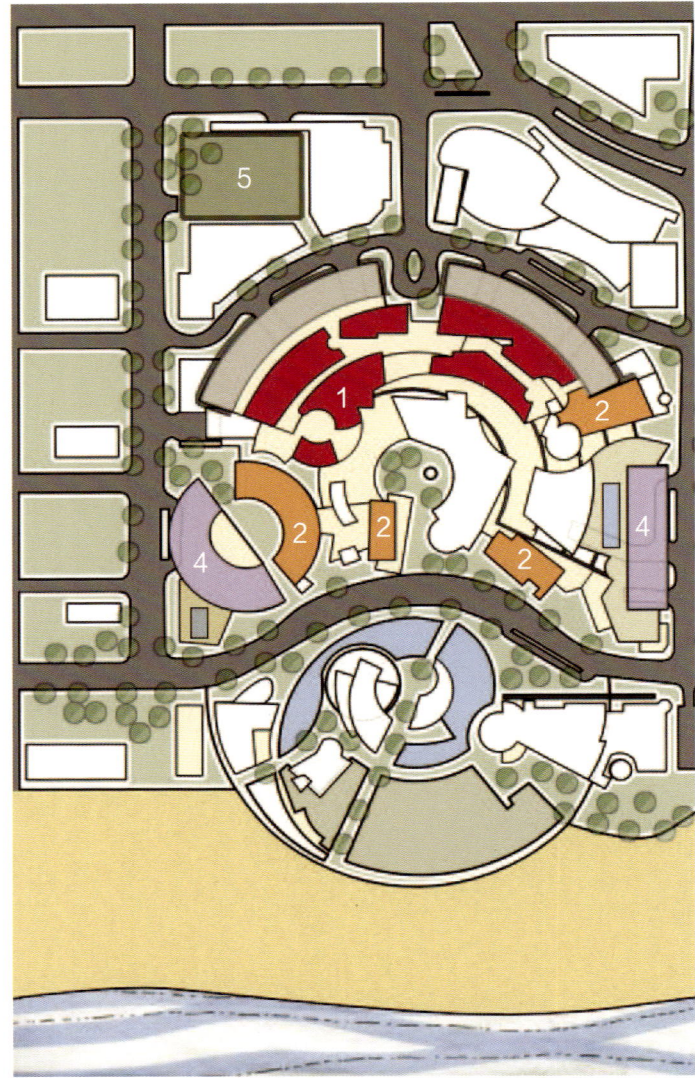

三层 Level 3

走廊 Corridor

圣彼得堡城市度假村
Urban Resort
Saint Petersburg

设计 willy muller Architects + BPG Arquitectos
地点 Saint Pitersburg, Russia
基地面积 135,000m²
建筑面积 15,925m²
总楼面面积 87,205m²
楼层 B3, 21FL
结构 Steel
饰面材料 Glass, Steel
甲方 GC Developement Group
结构工程师 BOMA
建筑工程师 THB Consulting

Design
Construction Engineer

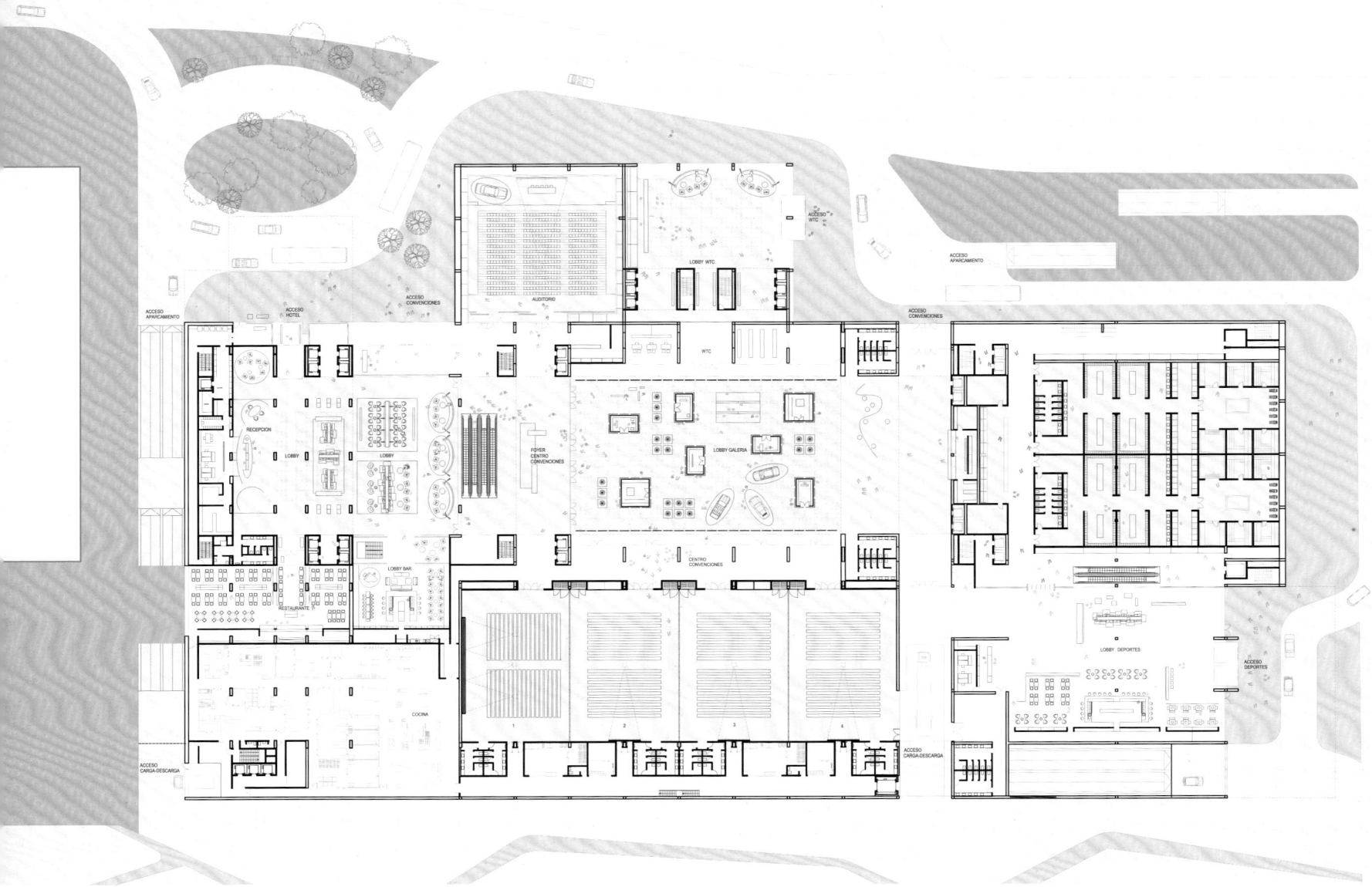

一层平面图 First floor plan

这座多功能综合体——城市度假村，被认为是一座独一无二的建筑，其在圣彼得堡使将来具有巨大发展前景的城市边缘地带各种不同的活动场所融为一体，轮廓鲜明。

仿佛从酒店表层削下来的"丝带"向不同方向弯曲起伏，勾勒出屋顶的形状，构成了体育运动场所及办公场所，并最终形成了整个建筑独特的外形。这座波状起伏的建筑外形不仅能在各条通道上一眼可见，而且使圣彼得堡这个原本规模较小的城市仿佛面积愈加增大了。高、中、低层建筑之间的共生关系促进了新建筑与周边环境之间外形上的过渡。"丝带"成了建筑的立面及屋顶，不断地折叠弯曲，最后形成运动场所及办公楼。

建筑最终的轮廓仿佛人工建造的景观，其外表可以积累积雪，并使积雪也成为建筑的材料，甚至成为项目的一部分。漫长的冬季里遍布城市各个角落的积雪覆盖了整个建筑的表层，从正面到屋顶，在天际线处勾勒出建筑更加清晰的轮廓，白色不透明的轮廓线与能够看到室外景色的透明外墙形成对比。

这座综合体包含一个设有 300 间房间的奢华五星级酒店，内设可容纳 500 人的会议中心及 23 间会议室及谈判室。另外还有一个功能齐全的运动中心，内设多种运动场地：可进行比赛及休闲娱乐的游泳池、体育馆、可跳舞和练习的舞厅、武术馆、壁球场乃至滑冰场。除酒店客人之外的外来人员也可进入所有这些供练习的空间和运动表演场地，但须经专门的通道进入，通道设有休闲及餐饮场所。在这些设施旁边是正在建设中的 24 000m^2 的办公楼及贸易中心，内设最先进的办公设备及大型的公司写字间。除此之外，该综合体还将建设一个构成五星级酒店的直升机停机坪，在酒店顶层将设一个米其林三星级饭店。

This multifunctional complex is thought as a unique building – an urban resort – that organizes different activities inside a volume with a recognizable profile on the city limits, a zone with a great development potential in St. Petersburg in the upcoming years.

This strong image is the result of different undulating movements of the "ribbons" that peel themselves off the volume of the hotel and divide the roof, organizing the athletic activities and the offices. This image of an undulating building is not only recognizable from the access roads, but also introduces an important element of scale, for the city of St. Petersburg in general has a low scale. This gradual symbiosis between the low, the medium and the tall volumes will help to create a formal transition between the context and the new building. The ribbons, which keep folding themselves until they transform into the sports and office building, will be treated like a facade and a roof at the same time.

The resulting profile expresses itself like an artificial landscape, that will incorporate the snow by means of surfaces that will accumulate the snow and make it into a material even make it part of the project. The snow, which can be considered an omnipresent element all over the city in the long winter months, will build up on the horizontal elements designed for that purpose on the entire surface from facade to roof, creating a contrast between these white opaque lines and the transparent surfaces through which the exterior can the observed.

The complex is composed of a luxurious five-star hotel with 300 rooms, equipped with a convention center with capacity for up to 500 people and 23 meeting and trading rooms. In addition there is a high yield sports center, with multiple fields for practicing all kinds of sports: Swimming pools for competition and leisure, a gym, ballrooms for dance and practice, martial arts dojos, squash courts and even an ice skating rink. All these spaces for practice and the performance of these sports have the possibility to receive external public, differentiating in designated circulations combined with an offer of leisure and catering. Alongside with these programs an office building and trade center of 24,000 square meters is being developed, equipped with the latest technologies for office management and large surfaces for firms. Apart from all these the complex also will have a Heliport and – forming part of the five-star hotel – there will be a Michelin three-star restaurant designed at the hotels top floor.

示意图 Diagram

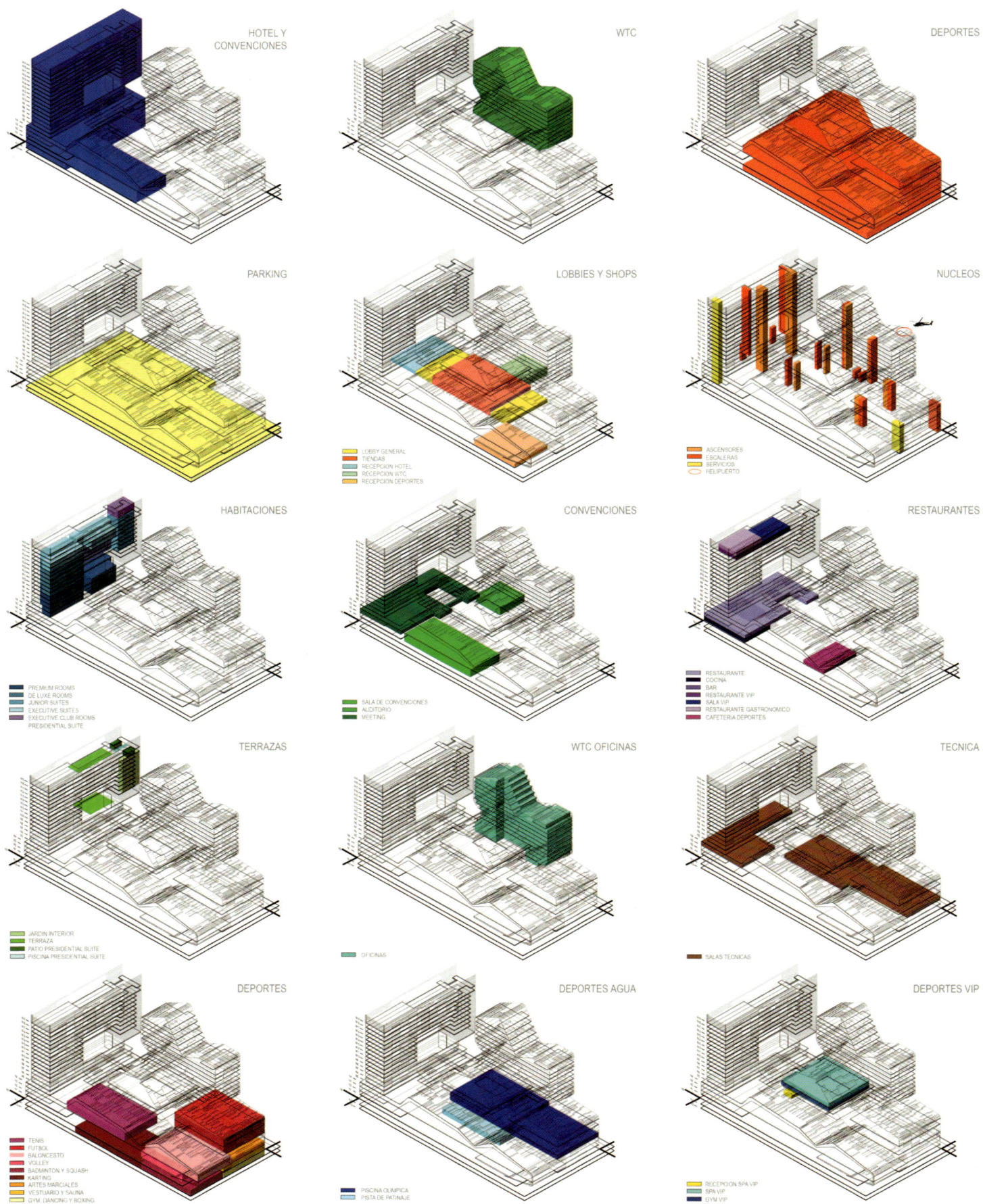

HOTEL Y CONVENCIONES

WTC

DEPORTES

PARKING

LOBBIES Y SHOPS

LOBBY GENERAL
TIENDAS
RECEPCION HOTEL
RECEPCION WTC
RECEPCION DEPORTES

NUCLEOS

ASCENSORES
ESCALERAS
SERVICIOS
HELIPUERTO

HABITACIONES

PREMIUM ROOMS
DE LUXE ROOMS
JUNIOR SUITES
EXECUTIVE SUITES
EXECUTIVE CLUB ROOMS
PRESIDENTIAL SUITE

CONVENCIONES

SALA DE CONVENCIONES
AUDITORIO
MEETING

RESTAURANTES

RESTAURANTE
COCINA
BAR
RESTAURANTE VIP
SALA VIP
RESTAURANTE GASTRONOMICO
CAFETERIA DEPORTES

TERRAZAS

JARDIN INTERIOR
TERRAZA
PATIO PRESIDENTIAL SUITE
PISCINA PRESIDENTIAL SUITE

WTC OFICINAS

OFICINAS

TECNICA

SALAS TECNICAS

DEPORTES

TENIS
FUTBOL
BALONCESTO
VOLLEY
BADMINTON Y SQUASH
KARTING
ARTES MARCIALES
VESTUARIO Y SAUNA
GYM, DANCING Y BOXING

DEPORTES AGUA

PISCINA OLIMPICA
PISTA DE PATINAJE

DEPORTES VIP

RECEPCION SPA VIP
SPA VIP
GYM VIP

剖面图 1 Section I

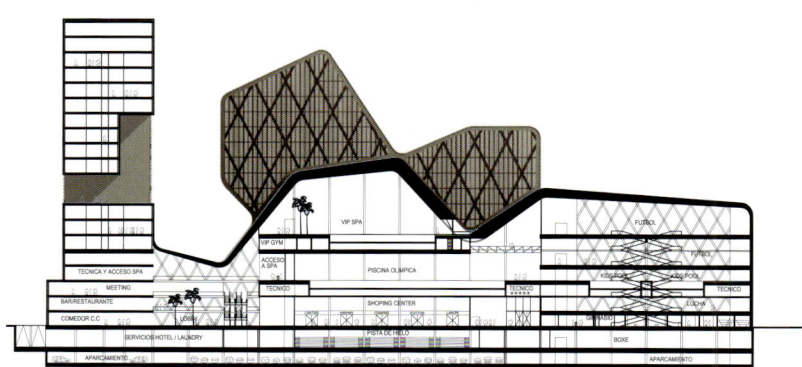

剖面图 2 Section II

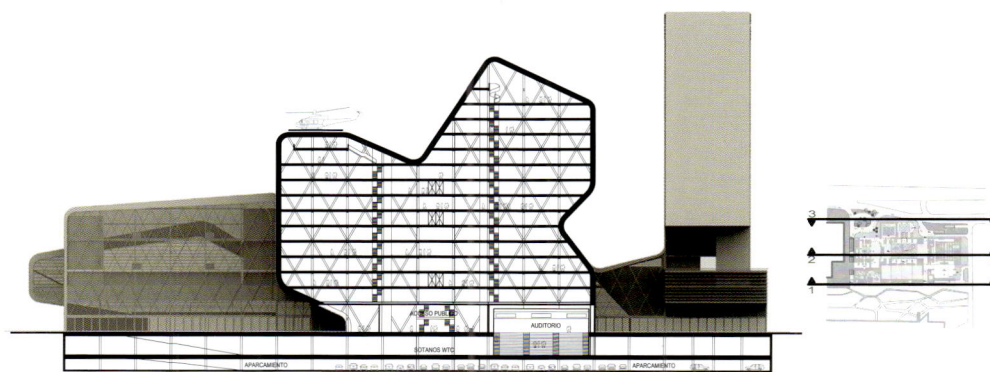

剖面图 3 Section III

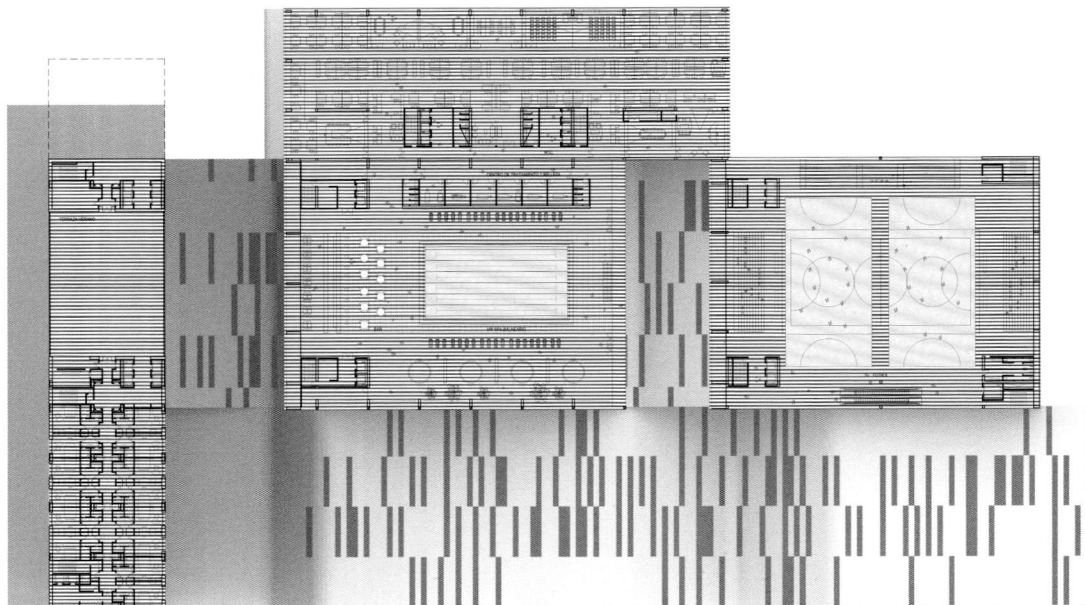

六层平面图 Sixth floor plan

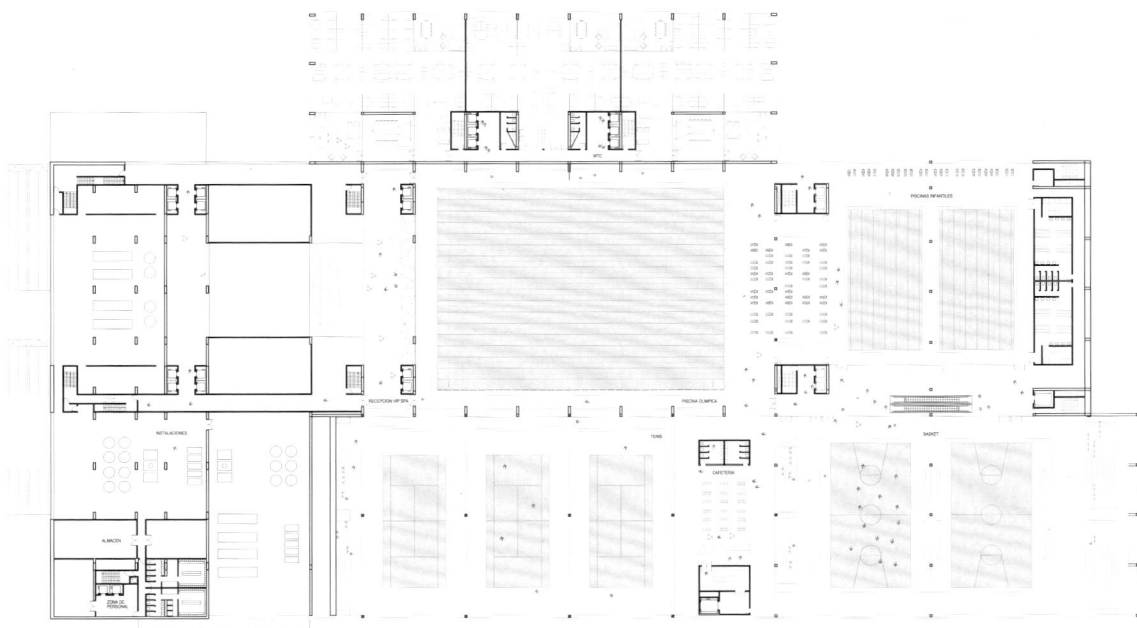

三层平面图 Third floor plan

Isla Moda 度假村
Isla Moda

地点　United Arab Emirates
总楼面面积　1,123,000m^2

总平面图 Site plan

鸟瞰图 Overall view

与世界上最著名的时尚设计师卡尔·拉格菲尔德及 KOR Hotel Group 合作，Isla Moda 度假村建于离迪拜海岸线 20km 外的人造岛屿上，其内设有 3 个宾馆及 150 栋别墅单元。设计的灵感来源于印度的漂浮宫殿及现代的等价交换——豪华游艇，Isla Moda 度假村通过全新的"人造"理念将其与"这个世界"上其他主题再生区分开，试图体验这个模拟的现实世界里的真实性。通过理解风格与时尚之间细微差别的本质，及它们与生命之间千丝万缕的联系，这个人工构造而成的岛屿采用造船技术建成。当小岛与这个有生命的环境融合在一起时，建筑与周围更加富有生气的自然环境之间的关系愈发和谐，从而使小岛达到其最佳状态。结构复杂的 320m×215m×24m 的大型体量——"黄金法则"的化身，最大限度地体现了各种体验与类型的多样化。

In collaboration with the world's most renowned fashion designer, Karl Lagerfeld and KOR Hotel Group, "Isla Moda" is a resort complex that comprises 3 hotels, and 150 residential units on a manmade island 20 km off the coast of Dubai. Inspired by the floating palaces of India, and the modern day equivalent – the cruise ship, "Isla Moda" distinguishes itself from other thematic reincarnations in "The World" by celebrating the notion of "manmade" and attempts authenticity within this simulacra of the real world. Utilizing ship building techniques, the fabricated island evolves from an intrinsic understanding of the nuances of style and fashion, and their inextricable link to the possibilities of how life can and should be experienced. When woven into the lived environment, it reaches an apogee through a symphonic balance between architecture and an elevated natural environment. A monolithic volume 320m×215m×24m – the physical embodiment of the Golden Rule, is intricately carved to establish maximum diversity of experience and typology.

外观 1 Exterior 1

示意图 Diagram

NATURAL LIGHTING
ON ALL UNITS

LIGHT COLOR ROOFING
& BALCONIES

CROSS VENTILATION

LED LIGHT FIXTURES

AUTOMATED SOLAR SHADES
OCCUPANCY LIGHTING CONTROL

PASSIVE SOLAR SHADING

NATURAL LIGHTING
ON ALL UNITS

BEACH VIEW

GARDEN
VIEW

1.5m

COOLING BREEZE FROM THE GARDEN

BREEZE FROM THE OCEAN

DEEP OVERHANGS

INSULATED LOW E GLASS

BUILDING INTEGRATED
PHOTOVOLTAICS

XERISCAPE PLANTS
(landscaping in ways that
do not require supplemental irrigation)

剖面图 Section

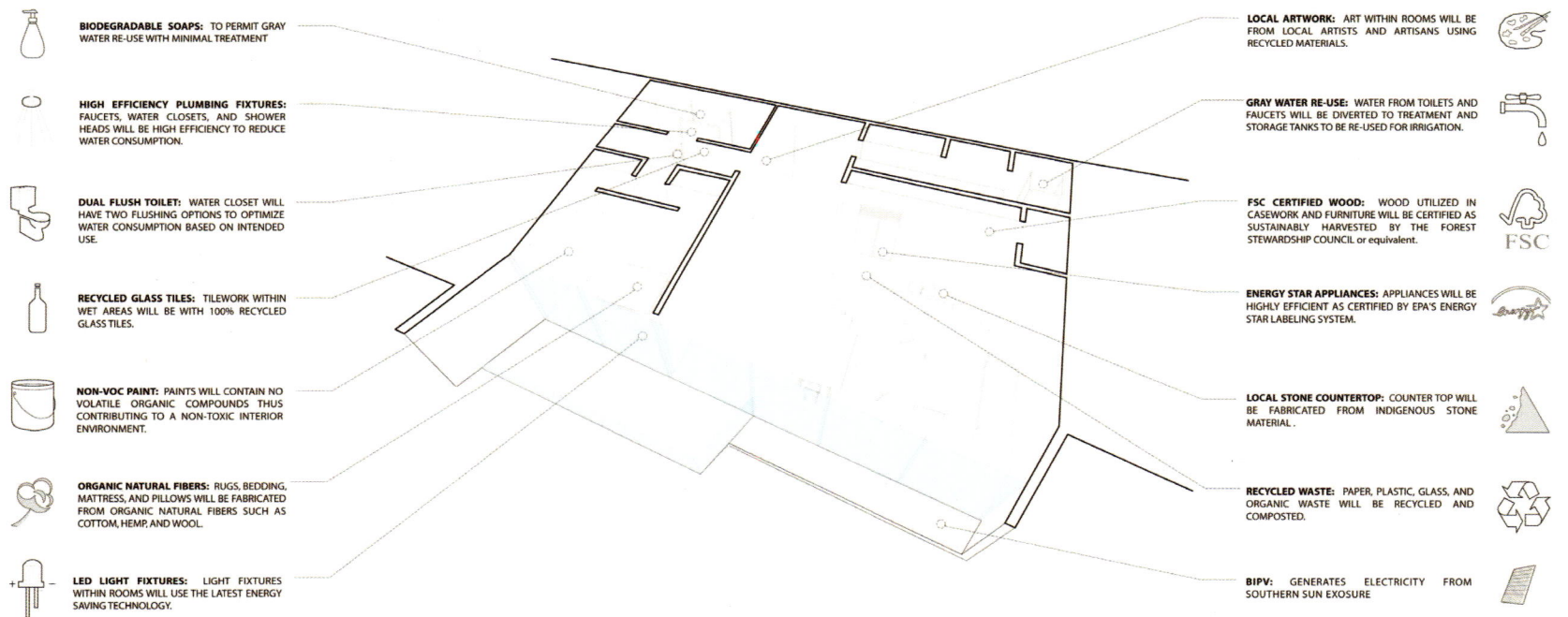

BIODEGRADABLE SOAPS: TO PERMIT GRAY
WATER RE-USE WITH MINIMAL TREATMENT

HIGH EFFICIENCY PLUMBING FIXTURES:
FAUCETS, WATER CLOSETS, AND SHOWER
HEADS WILL BE HIGH EFFICIENCY TO REDUCE
WATER CONSUMPTION.

DUAL FLUSH TOILET: WATER CLOSET WILL
HAVE TWO FLUSHING OPTIONS TO OPTIMIZE
WATER CONSUMPTION BASED ON INTENDED
USE.

RECYCLED GLASS TILES: TILEWORK WITHIN
WET AREAS WILL BE WITH 100% RECYCLED
GLASS TILES.

NON-VOC PAINT: PAINTS WILL CONTAIN NO
VOLATILE ORGANIC COMPOUNDS THUS
CONTRIBUTING TO A NON-TOXIC INTERIOR
ENVIRONMENT.

ORGANIC NATURAL FIBERS: RUGS, BEDDING,
MATTRESS, AND PILLOWS WILL BE FABRICATED
FROM ORGANIC NATURAL FIBERS SUCH AS
COTTON, HEMP, AND WOOL.

LED LIGHT FIXTURES: LIGHT FIXTURES
WITHIN ROOMS WILL USE THE LATEST ENERGY
SAVING TECHNOLOGY.

LOCAL ARTWORK: ART WITHIN ROOMS WILL BE
FROM LOCAL ARTISTS AND ARTISANS USING
RECYCLED MATERIALS.

GRAY WATER RE-USE: WATER FROM TOILETS AND
FAUCETS WILL BE DIVERTED TO TREATMENT AND
STORAGE TANKS TO BE RE-USED FOR IRRIGATION.

FSC CERTIFIED WOOD: WOOD UTILIZED IN
CASEWORK AND FURNITURE WILL BE CERTIFIED AS
SUSTAINABLY HARVESTED BY THE FOREST
STEWARDSHIP COUNCIL or equivalent.

FSC

ENERGY STAR APPLIANCES: APPLIANCES WILL BE
HIGHLY EFFICIENT AS CERTIFIED BY EPA'S ENERGY
STAR LABELING SYSTEM.

LOCAL STONE COUNTERTOP: COUNTER TOP WILL
BE FABRICATED FROM INDIGENOUS STONE
MATERIAL .

RECYCLED WASTE: PAPER, PLASTIC, GLASS, AND
ORGANIC WASTE WILL BE RECYCLED AND
COMPOSTED.

BIPV: GENERATES ELECTRICITY FROM
SOUTHERN SUN EXOSURE

绿化单元示意图 Green unit diagram

外观 2 Exterior II

立面图 Elevation

外观 3. Exterior III

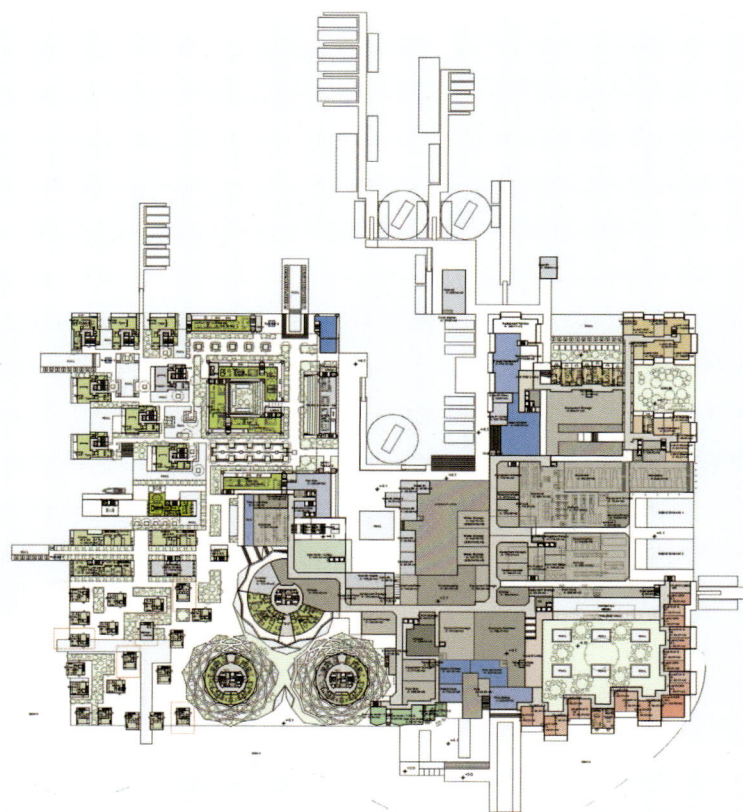

二层平面图 Second floor plan

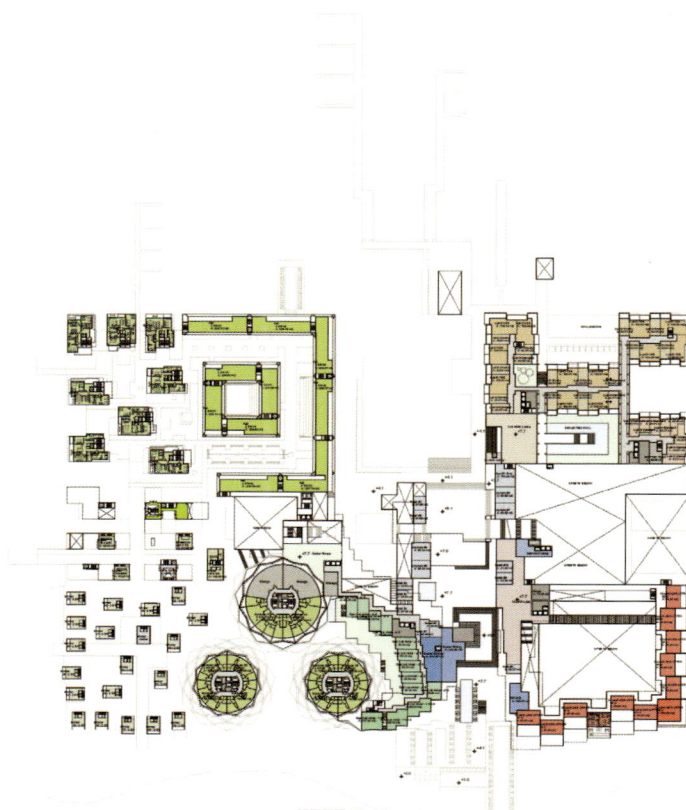

一层平面图 First floor plan

外观 4 Exterior IV

剖面图 1 Section I

剖面图 2 Section II

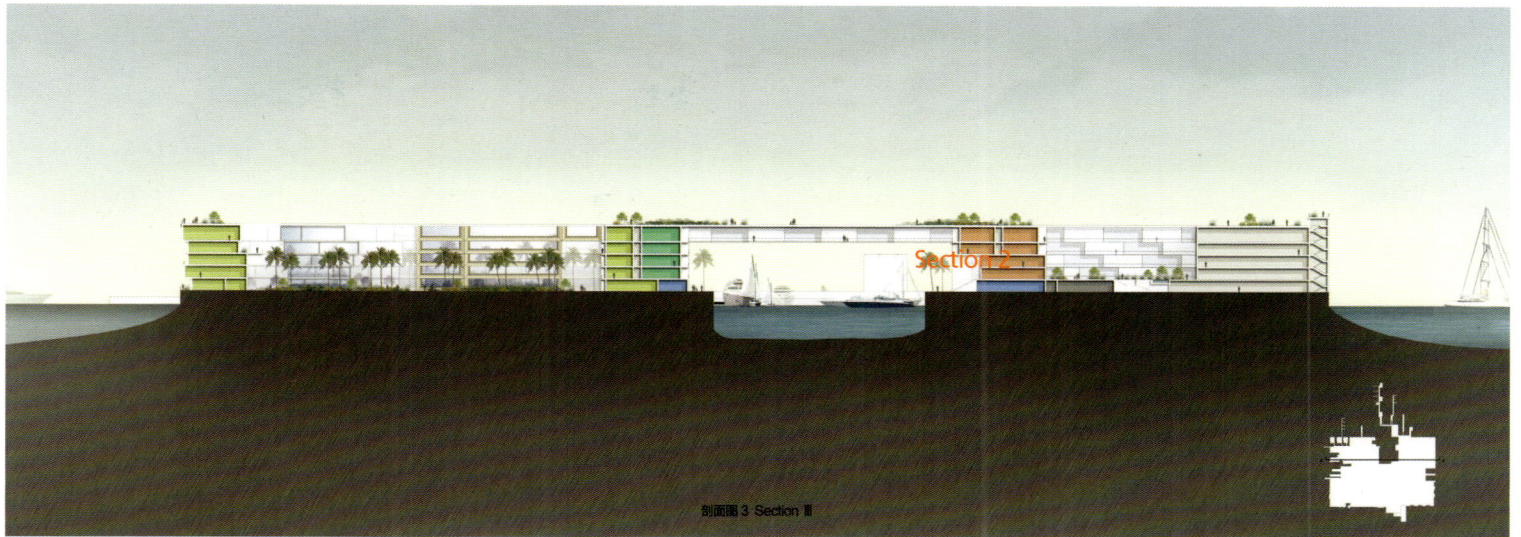

剖面圖 3 Section III

剖面圖 4 Section IV

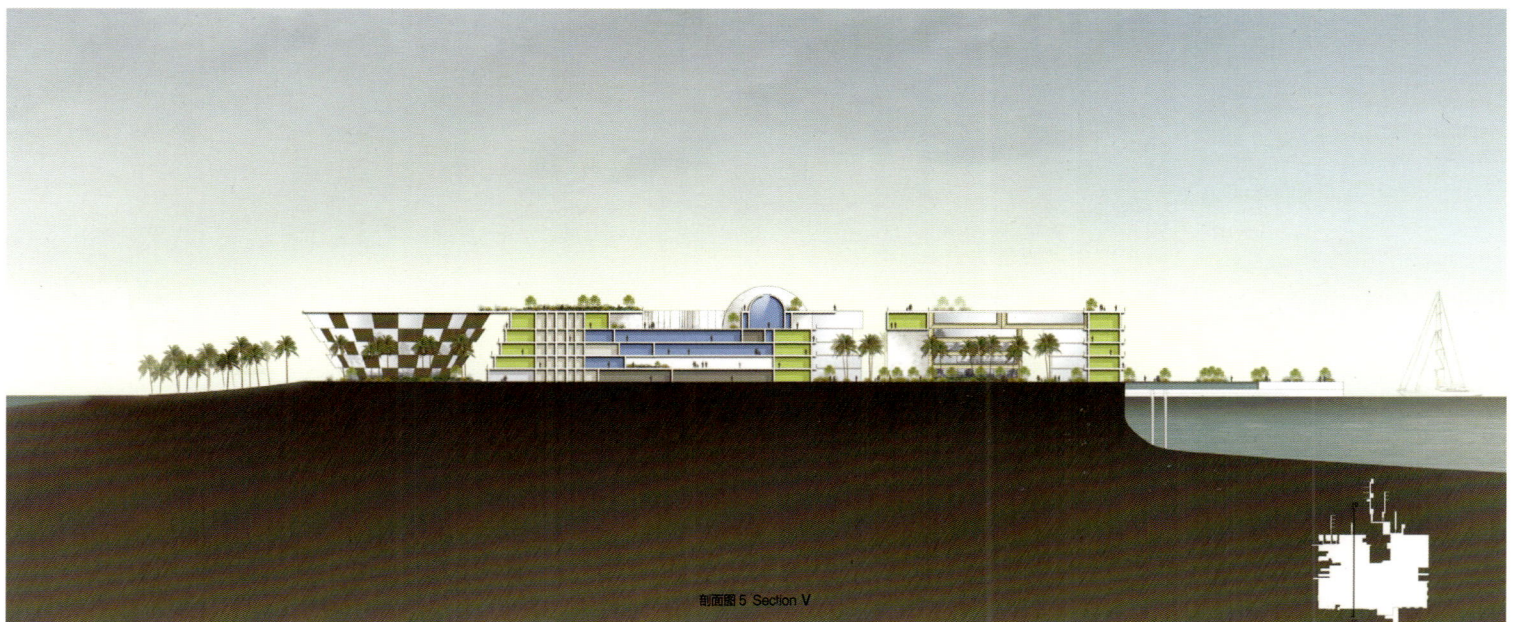

剖面圖 5 Section V

墨西哥度假村
Mexico Resort

设计 LAVA / Alexander Rieck, Chris Bosse,
地点 Tobias WallisserPacific coast of Mexico
甲方 Confidential
参与设计人员 Wenzel+Wenzel Stuttgart

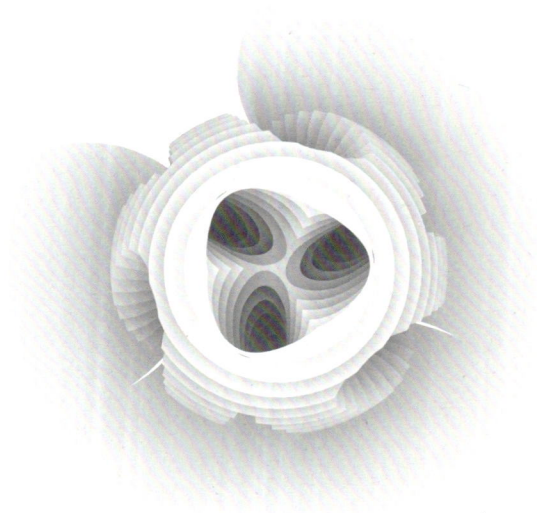

电子模型 1 Digital model 1

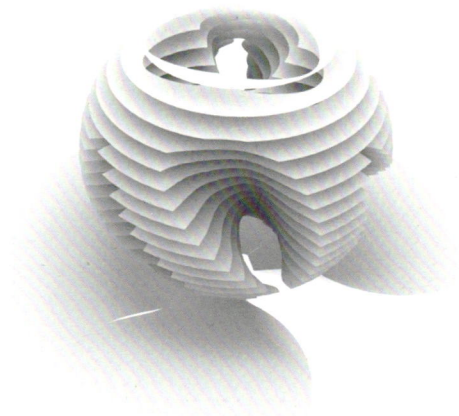

电子模型 2 Digital model 2

全景 1 Overall view 1

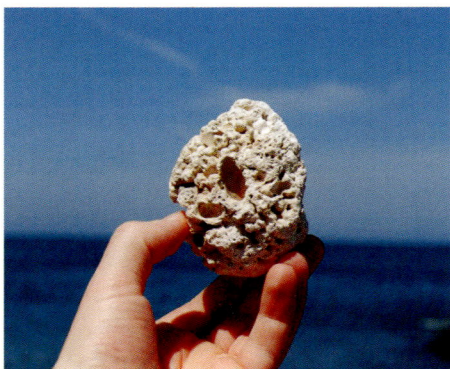

来源于当地的设计灵感 Local inspiration

全世界的旅游胜地都拥有共同的特征，游客都追求旅游景点的怡人风景、惊险刺激及当地人们热情好客的礼待。

为了赋予旅游胜地独有的特征，在自成一体的度假村内具有当地某一特性是非常必要的。地方艺术、色彩、植被、动物群等都能很好地引导该地区建筑的设计。

对"奢侈"一词现在的定义在将来会发生转变，不再是限量的物品，而会指身心的健康长寿，令人们重新发现人与自然之间紧密的联系。"绿色"将成为新的黄金标准。

健康与可持续的生活方式将取代对能量消耗的需求。如今的旅游景点消耗了过量的能量及自然资源。

因此，设计师的目标是开发一个可持续的度假村，为节能旅游设定新的标准。基于以零碳及零垃圾为特点的零排放原则，墨西哥度假村将同时保持生活的最优品质与"奢侈"。

墨西哥度假村的开发依靠可再生能源，这将提高人们的社会责任感及对自然环境的保护意识。

通过运用智能建筑技术，如交互式遮阳设施、自然采光、通风系统以及当地原始材料的使用，确保了度假村能够满足可持续性的要求。

Worldwide tourist destinations have common features. The customer searches for beauty, adventure, and hospitality.
In order to give a destination a unique character, it is necessary to provide a local identity within the self-contained resort. Local art, colours, vegetation and fauna would ideally guide the design of this location.

Current "luxury" definitions will be shifted in the future from exclusive items towards wellness and longevity of the body and mind, creating a rediscovery of our deep bond with nature. Green will be the new gold.

A lifestyle of health and sustainability will replace the need for consumption. Current tourist attractions consume excessive amounts of energy and natural resources.

The goal here is to develop a sustainable resort, setting new standards for energy efficient tourism. Constructed on Zero Emission principles characterized by zero carbon and zero waste, the resort will simultaneously maintain the highest quality of life and luxury.

Relying on renewable energy resources, this development will promote social responsibility and the protection of the natural environment.

Implementing intelligent building technologies such as interactive shading devices, natural day lighting and ventilation systems, as well as the use of locally sourced materials, ensures that these targets will be met.

瓦地度假村
Wadi Resort

设计　Chad Oppenheim
地点　Wadi Rum, Jordan
基地面积　80,000㎡
结构工程师　Oppenheim Architecture

抵达之处 Arrival

国际知名的绿色建筑公司乍得 Oppenheim 建筑设计院完成了一个标志性的生态环保设计方案——瓦地度假村。该度假村位于约旦瓦地伦，预计将于 2014 年竣工。Oppenheim 建筑设计院（OAD）在国际竞标中赢得了瓦地度假村的设计权，随后，他们将开启一个规模宏大的度假村项目的建设，该项目将建造 47 个沙漠度假旅店，其会为环球旅行爱好者提供一次非凡的原始住宅体验。瓦地度假村距佩特拉市仅一个半小时的路程（佩特拉市——著名的纳巴泰人古城，它的建筑大都雕刻在沙漠岩石之上）。

Oppenheimd 的中标方案致力于用一种不同于社会上普遍的自然建筑开发方式来重新诠释这个度假村，即综合考虑这里的自然特色———一个沙漠与峭壁相汇处的神奇峡谷。度假村很自然地与周围壮观的景色融为一体，开拓并丰富了沙漠景观的天然之美，为游客提供了极具原始风情又不失奢华感的沙漠旅店。这些旅店和公馆的位置醒目而险峻，每间都有自己独特的形态，但是它们又都倾吐着当地的文化，也与这里的地形浑然天成。所以，旅店最终的成品也是对华丽建筑的全新演绎：建筑设计虽然华丽，但是这种华丽却毫不冗余，恰如其分。

旅店的框架结构嵌入沙石峭壁，而旅店的形状则依据岩石的形状自然形成。其他的结构则使用夯土、水泥和当地的红色沙子混合建造。尽管这些建筑的形态变化很少，但是这些形态的作用却很大：不管是建造的还是雕刻的旅店，都努力去使建筑能够协调和平衡，同时又能形成和丰富周围的景观。建筑师还特意模糊了这些旅店的室内和室外景观，这样，通过最少的工作却达到了最大的效果。Oppenheim 建筑设计院的建筑师们从原始的山洞房屋中获得灵感，同时运用自己在环保设计上的专业技术，为度假旅店创造了被动式十字通风系统，这样便能充分利用岩石的自然冷却效应；而且恰当的定位使建筑的能量消耗达到最小化，同时又使人们的舒适健康生活达到最优化。

"我们已经练就好了我们的各种感官去看、去闻、去品味、去倾听、去感受瓦地伦的神奇之美。所以，通过原始的、本能的设计冲动，同时从自然的力量、节奏和形态（即过去、现在和将来）中得到灵感，我们发掘出了沙漠中最原始的能量。"在谈及这项工程的创作过程时 Oppenheim 提到。

此项工程所使用的技术都是几千年前曾经使用过的。我们从那些生活在这片美丽而神奇的沙漠地区几千年的人们那里学习到很多。我们特意使用了当地的材料，也采取了各种节水措施（节省居民饮用水和灌溉用水），目的是在地下蓄水池中建造一个相对封闭的雨水收集系统，并通过"活机器"——植物、动物等自然生物，来收集黑水和灰水。（注：灰水主要指厨房用水、沐浴用水和清洗水等；黑水指尿、粪未经分离的冲厕水。）所有这些环保系统和环保服务都融入到了我们的设计中。

这个占地 80 000m^2 的建筑项目响应了当地宝贵的地区特色：一种建筑风格，经过千年的发展完善，终于又以一种清晰、恰当的姿态出现在中东地区。

International "green" architect Chad Oppenheim sets a new benchmark for design and ecological sensitivity with the Wadi Resort – located in Wadi Rum, Jordan, set for completion in 2014. Oppenheim Architecture + Design (OAD) beat out a global competition and will execute an unprecedented project comprised of 47 desert lodges, setting forth a future primitive experience for the avid globetrotter, an hour and a half outside of Petra, the ancient city of the Nebataeans carved into the desert rock.

Oppenheim's winning proposal set out to reinterpret the way society deals with surrounding nature by taking full advantage of the mystical valley where desert sand meets desert stones. The project merges silently with its wondrous setting, exploiting and enhancing the natural

旅店 1 Lodge I

旅店 2 Lodge II

全景 Overall view

beauty of the desert to establish accommodations that are uniquely elemental and luxurious. Dramatically situated, the lodges and villas in their various incarnations are all about a visceral connection to culture and place. The resulting experience is a revolutionary notion of opulence that is intentionally reduced to what is essential.

The structure of the lodges will be carved into the sandstone cliffs, utilizing the existing geological geometries of the rock to devise the form. Other structures are comprised of rammed earth and cement mixed with the local red sand. The minimal yet powerful gestures of the architecture, both built and carved serve to create harmony and balance, while framing and amplifying the surroundings. The interior and exterior are deliberately blurred establishing maximum impact with minimum effort. Inspired by the primordial, Oppenheim used his expertise in sustainable design to create passive means of cross ventilation, taking full advantage of the natural cooling effect of the rocks, and proper positioning allowing the project to minimize energy consumption and maximize comfortable healthy living.

"We have trained and heightened our senses to see, smell, taste, hear, and touch the mystical beauty of Wadi Rum. We tapped the inherent power of the desert through primal and instinctual design moves, informed by the forces, rhythms and patterns of nature – past, present, and future," says Oppenheim about his creative process for the project.

The strategies employed are those that have been proven over the last thousands of years. We have learned a great deal from the civilizations that have lived in the beautiful and magical desert for millennia. Great care has been given to utilizing local materials as well as various water conservation measures for both human and site irrigation to establish a relatively closed system of harvesting rain water in subterranean cisterns and re-harvesting grey/black water through a living machine of botanical and biological nature. All systems and services will be completely integral to the design.

The 80,000 ㎡ architectonic form responds directly to the rich regional cues: an evolutionary process that has established, over millennia, a clear and appropriate identity found in the Middle East.

透视 4 Lodge IV

安宁河水上休闲度假中心
Anning River Aquatic Recreation Resort

设计　Studio Shift
地点　Sichuan Province, China
基地面积　47,000m^2
建筑面积　23,500m^2
总楼面面积　35,000m^2
楼层　2FL(aquatic center), 10FL(residential tower)
结构　Steel, Cast-in-Place Concrete
饰面材料　Perforated metal building panels, Landscaped roof,
Board-formed concrete
设计团队　Irina Krusteva, Stephen Morton(Studio Shift),
Gerdo Aquino, Ying Yu Hung, Dawn Dyer(SWA
Group)
甲方　Government of Miyi County
景观设计　SWA Group
结构工程师　Thornton Tomasetti

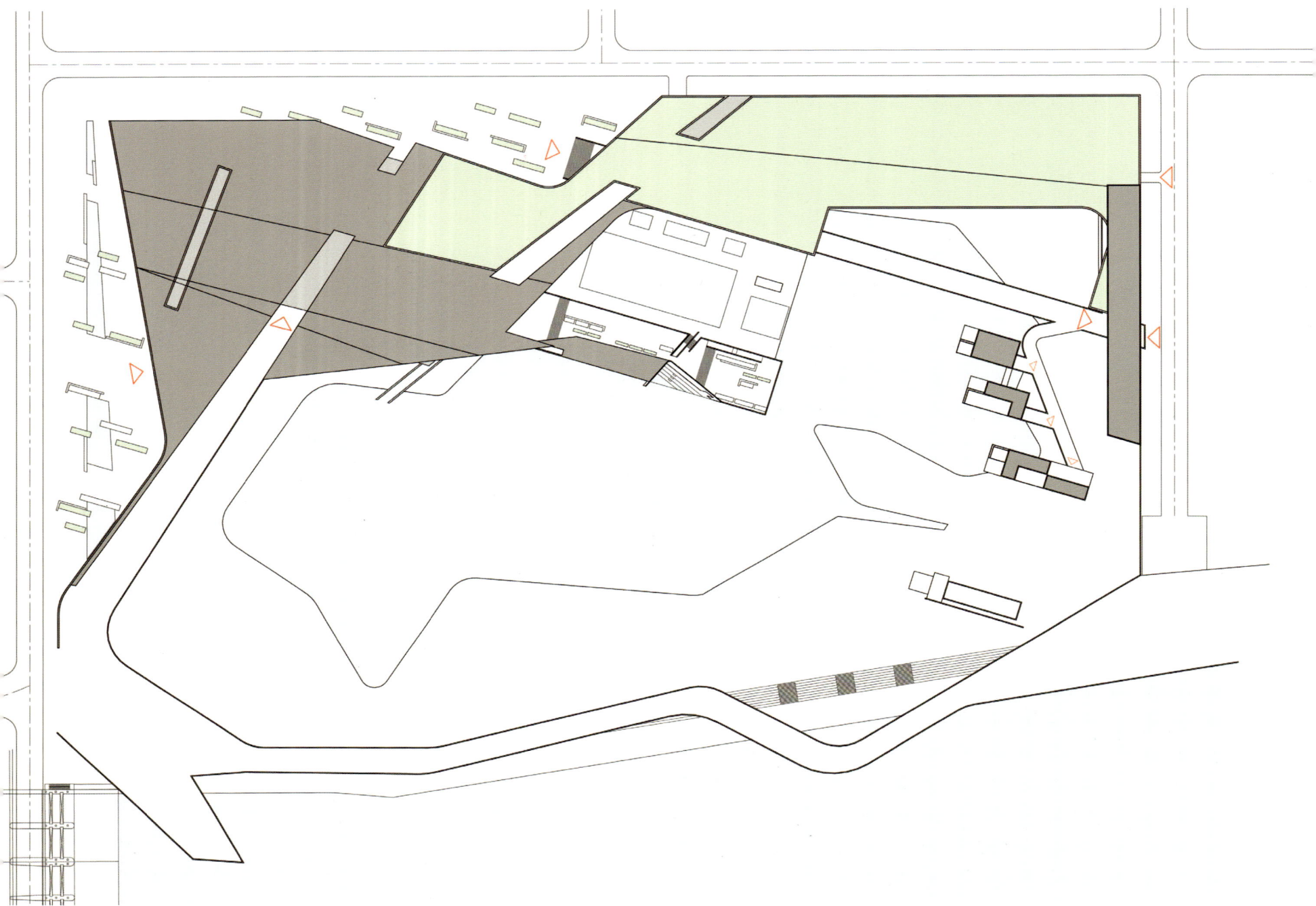

总平面图 Site plan

立面图 Elevation

在创造出一个家庭和水上休闲中心新模式的同时，该全新的水上中心也重新定义了城市便利设施所扮演的公众角色。通过将河边堤岸和沿水电站堤坝的人行道直接铺设到整个大厅，该中心尽力将所有邻近的公用设施连接在了一起。位于上层的通风体系贯穿整个建筑，将这些各自分离的项目元素和场地边界连接起来，并最终将整个结构系统地勾勒出来。在这个水上世界里，通过交叉和重叠，场地里不同的元素和轴线得以调和，创造出混合交错的时间和空间感，给项目增加了一个多孔的边界，而该项目之前是趋于保守并与城市生活区别开来的。

室外建造的两个公共区域用于水上休闲活动，更加强烈地突出了该中心的目的性，同时也降低公众对设施的使用压力。这些公共区域可用作大型集会场所，为众多的集会提供场地，同时也适用于休闲聚会。为了进一步鼓励在该中心进行公众活动，零售商店位于该中心东边，与街道水平，但低于停车场，目的就是为了使该设施拥有多样化的用途。在该中心上部的环形体系内这些零售商店被直接复制，作为过渡项目，连接起了水上设施的核心部分和南边的住宅区。

为了四季均可招徕游客并满足游客需求，该水上中心拥有全套室内游泳设施，包括游泳池和潜水池、不同性质和温度的休闲池，还有适合小孩子的嬉戏池和全套的露天温泉设施。游客和水可以全方位亲密接触，亲近自然。该温泉部分没入场地下面，与冲浪池的联系将会得到进一步加强。该中心的室外便利设施也一应俱全。高处的阳台和与水池同一平面的凉亭环绕着一系列外部游泳池，供游客放松身心。配套的更衣室、户外的咖啡厅和饭店更加强了这一愉悦体验。该中心南侧是一栋住宅楼，可提供 80 间客房和 6 间建于水畔的特殊开放式套房，创造出一种与水池的独特关联。住宅楼和邻近规划的背景在用途和面积上相呼应，又在游泳池南部边缘创造出一个半私密的空间。

The new Aquatic Center simultaneously creates a new model for family and water-based recreational tourism while redefining the public role of such an amenity within the city. The Center immediately strives to connect to all adjacent public infrastructures by bringing the riverfront promenade and the pedestrian movement alongside the hydroelectric dam directly into the lobby. This circulation system which penetrates the building at the upper level forms the connective tissue linking the disparate program elements and site edges, ultimately, delineating the overall programmatic organization of the structure. Here various site forces and axes are mediated through intersection and overlapping, creating hybridized moments and spatial conditions that bring a rather porous edge condition to a program that conventionally assumes an introverted posture and shelters itself from the urban condition.

The creation of two significant exterior public spaces programmed with casual water activities alludes to the more intense purpose served by the Center and allows the public to engage the project in a more passive capacity. These public spaces assume a dimension appropriate to large-scale aggregation in order to facilitate a broad range of high-intensity activities as well as casual gathering. To further enhance the public life of the Center, retail space is provided on the eastern edge of the project at street level, below the parking garage, intending to create a diversification of program and usage types. The retail program is replicated directly above as a transitional program on the upper circulation system within the Center bridging between the heart of the aquatic facility and the residential program on the southern edge.

In order to promote and accommodate year round tourism, the Aquatic Center features complete indoor pool facilities including swimming and diving pools, leisure pools of varying natures and temperatures, play pools for children and complete on-site spa facilities. The connection to water throughout is intimate and of a direct nature. The spa level descends partially into the site where its relationship to the swimming lagoon is further enhanced and differentiated. The Center's outdoor amenities are just as comprehensive with a series of exterior pools surrounded by a raised sun deck and pool level cabanas for relaxation. Associated changing facilities and an open-air café and restaurant further augment the experience. Anchoring the southern edge of the Center is a residential tower featuring over 80 rental units and 6 special suites that are sited directly into the water, creating a unique connection to the lagoon. The residential tower responds to the adjacent, planned context in usage and scale while simultaneously creating a semi-private condition for the southern edge of the swimming lagoon.

水上中心项目示意图 Aquatic center program diagram

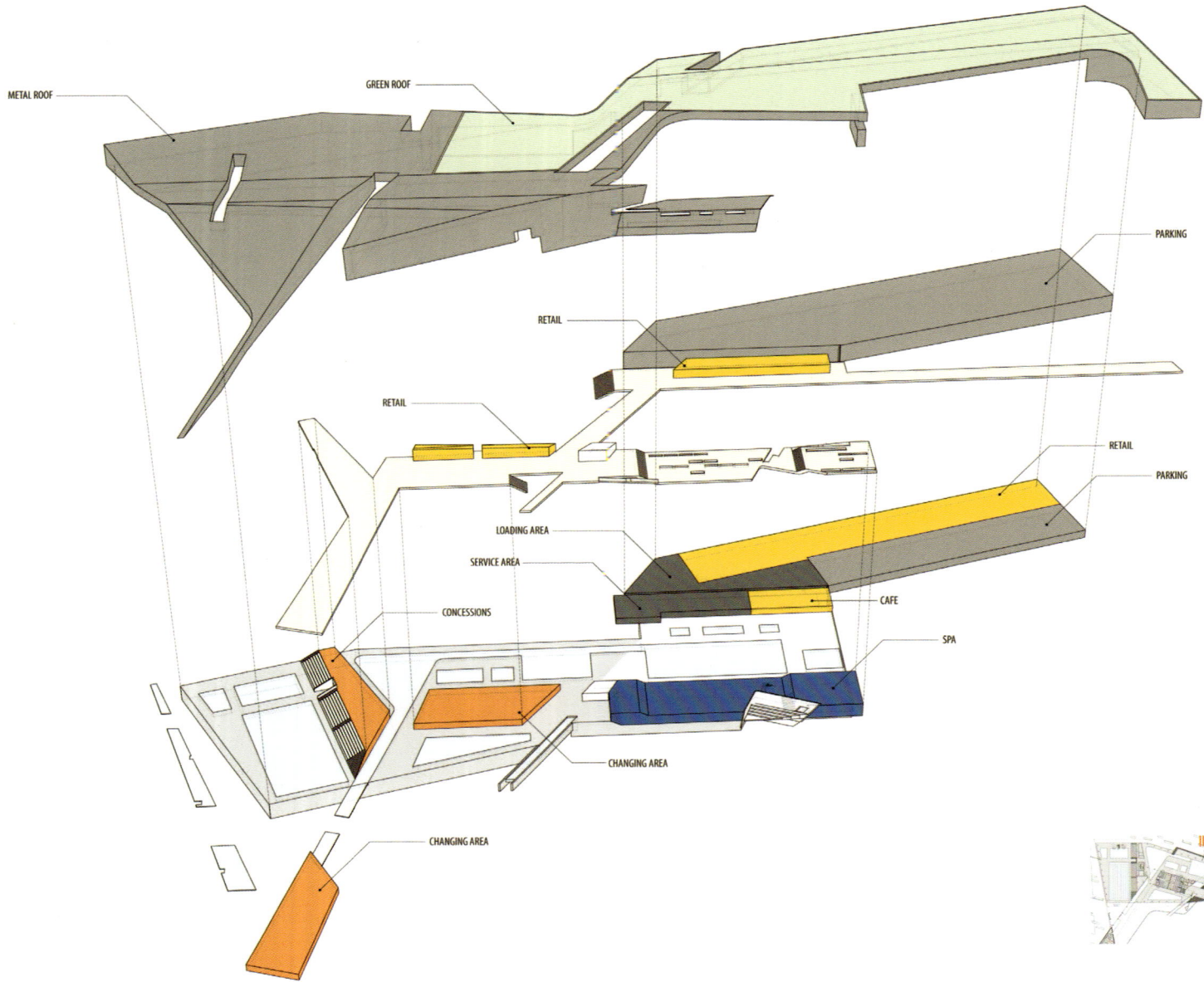

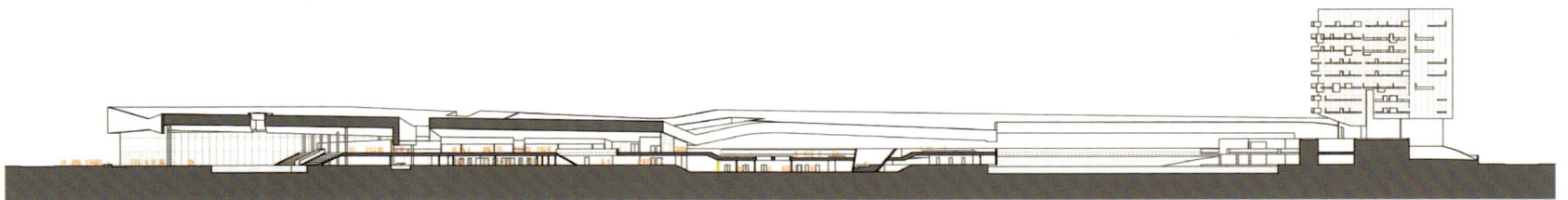

METAL ROOF

GREEN ROOF

PARKING

RETAIL

RETAIL

RETAIL

PARKING

LOADING AREA

SERVICE AREA

CAFE

CONCESSIONS

SPA

CHANGING AREA

CHANGING AREA

剖面图 1 Section I

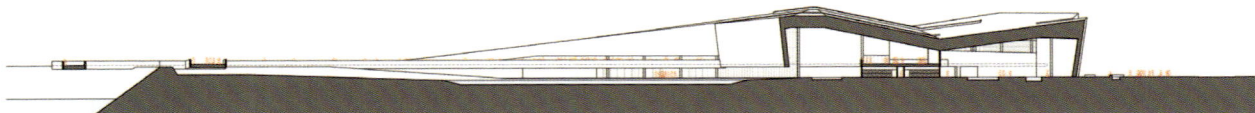

剖面图 2 Section II

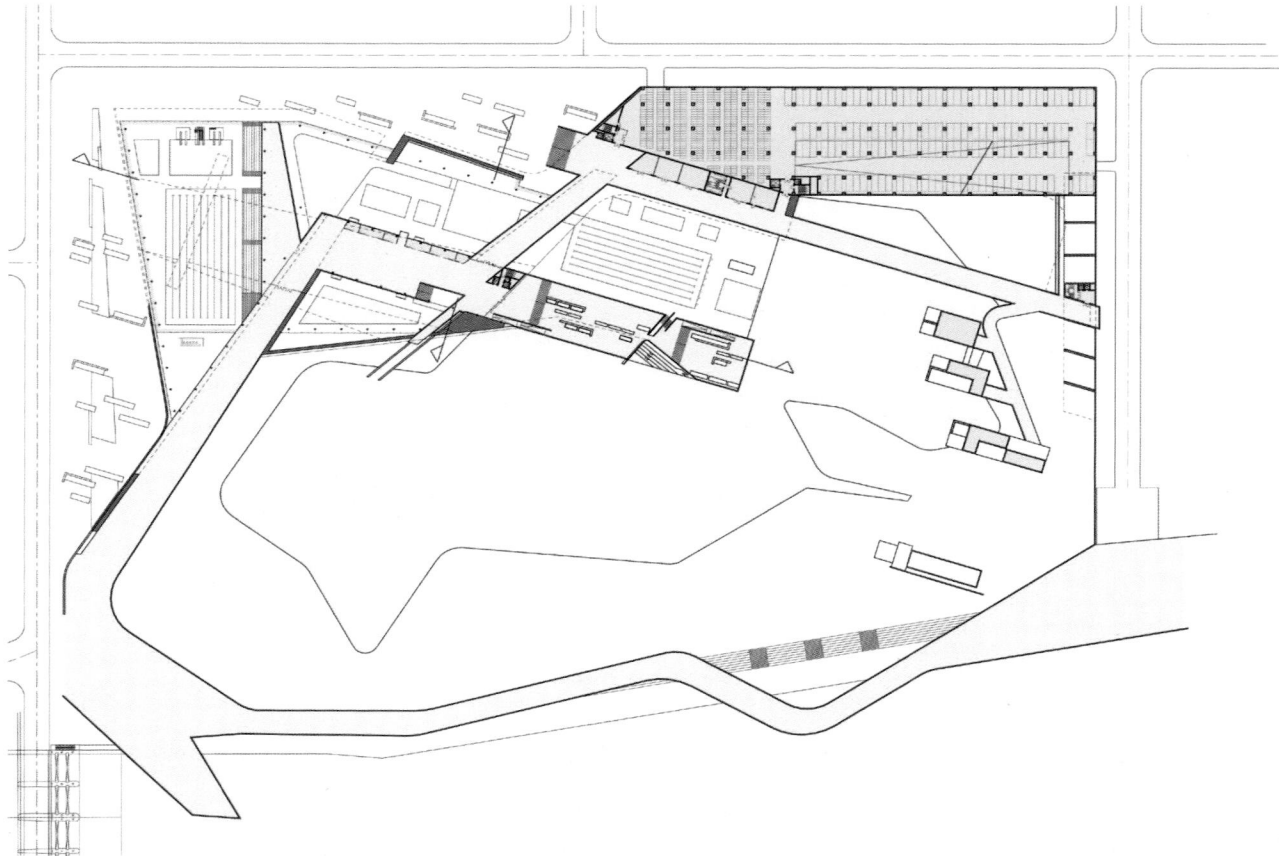

二层平面图 Second floor plan

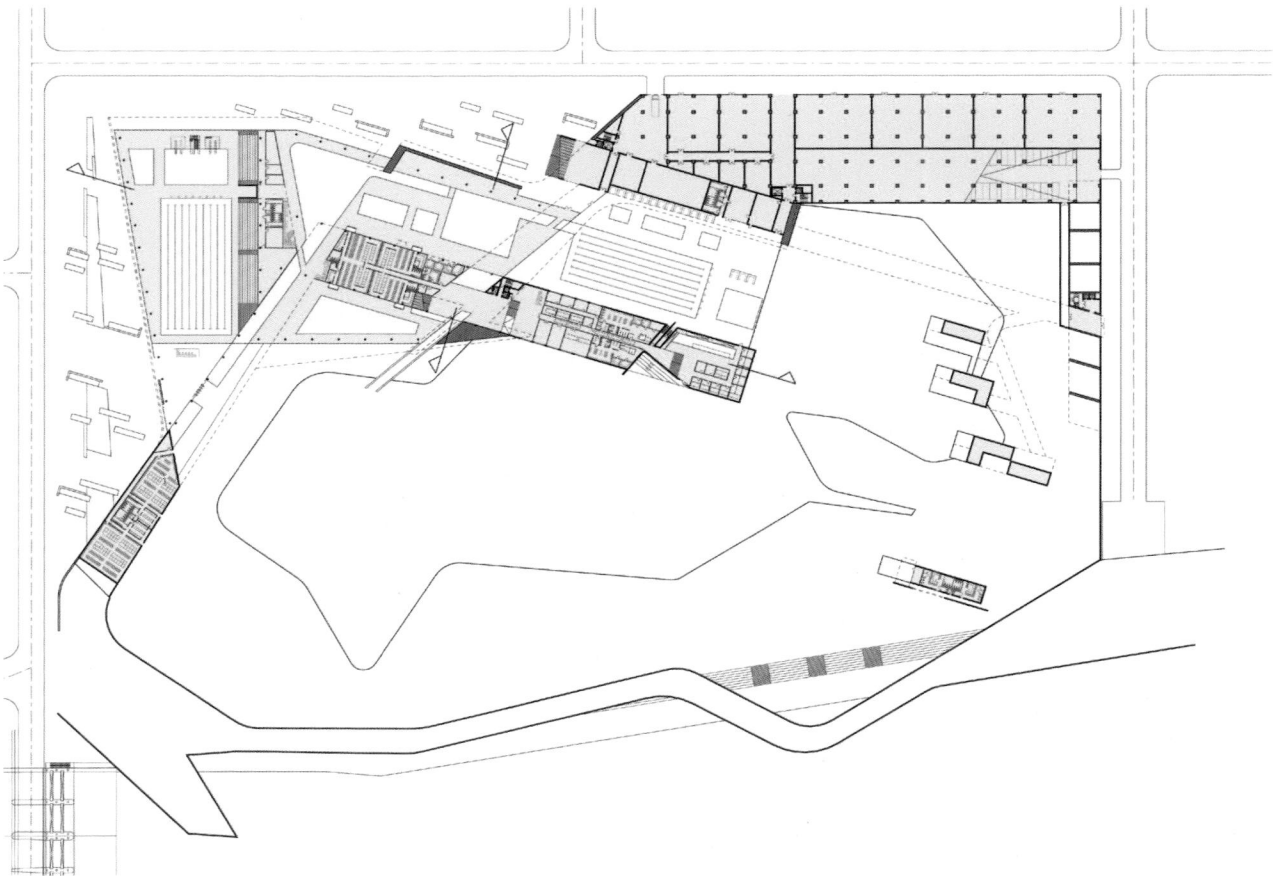

一层平面图 First floor plan

伦敦 W 酒店 W Hotel London

酒店 HOTEL

圆形别墅
Villa Ronde

地点　Japan
设计　Henri Gueydan & Fumiko Kaneko + Ciel Rouge Creation
建筑面积　1,800m^2
饰面材料　Wood flooring, Concrete wall, Board walls
摄影师　Kitano kensetsu
甲方　Iishi. T

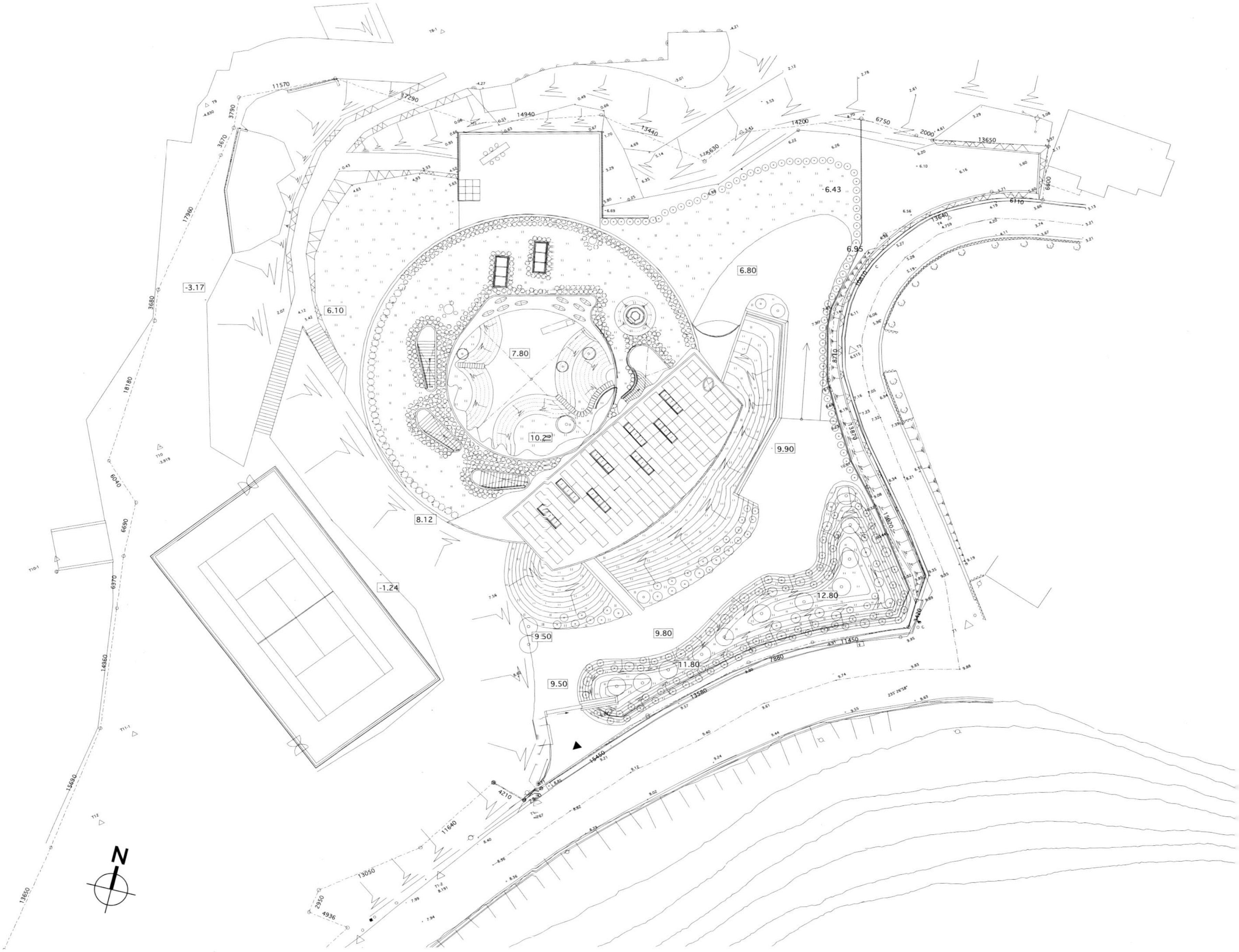

N

-3.17

6.10

7.80

8.12

10.2

-1.24

6.43

6.80

6.95

9.90

12.80

9.50

9.80

11.80

9.50

总平面图 Site plan

这座别墅建筑位于日本海岸边，包括一个博物馆、一间客房以及一个度假屋。别墅空间的组织布局非常自由，可以将房间封闭起来，也可以打开房间形成一个围绕着天井的连续空间。圆形是能让人看到最多周围美丽风景的形状，不仅如此，它还能抵御强劲的台风。建筑本身看起来像是从一座小山上生长出来的，山上的新鲜空气循环往复，可以为别墅提供永久性通风。建筑师用尽一切办法来保证建筑的自然保温隔热效果：立面设计成双层的，可以防止强风侵袭、骄阳曝晒；屋顶上还覆盖了 30cm 厚的泥土以及一个洒水系统。建筑的颜色与岩石相同，好像是从绿色的植物中凸现的一块岩石。在建筑内部，所有房间彼此相连，犹如一个大型美术馆，又像一间蜿蜒曲折的住宅。看起来好似孔洞的窗户不但能聚焦重要的景色，还能保证空间的私密性。建筑师 Ciel Rouge 的创作基于东西方文化的融合，并将这种理念运用到了工程与项目中。

By the Japanese coast, this building includes a private museum, a guest house and a resort. It is thought as a wide free organic space in which rooms can be closed or in continuity to each other around a patio. The round shape is the best to cover the beautiful view around as well as to resist and glide in the strong typhoon winds. The building itself seems to grow from an hill in which air system circulates to ventilate permanently the house. Everything is thought for the best thermic natural ratio with a double facade for protecting from winds and sun, as well as the roof is covered by 30cm of earth including a watering system. The building takes the same color from the rocks as it was emerging in the green. Inside it is thought as all the rooms are connecting with each other's and making like a huge gallery or a wandering house. Windows look like holes used to focus the important points of the landscape and to keep the space privacy. Ciel Rouge creation is engaged in projects and achievements based on a blend of cultures.

从屋顶看室外 Outside view from rooftop

从楼梯看天空 Skyview from staircase

室内 1 Interior I

室内 2 Interior II

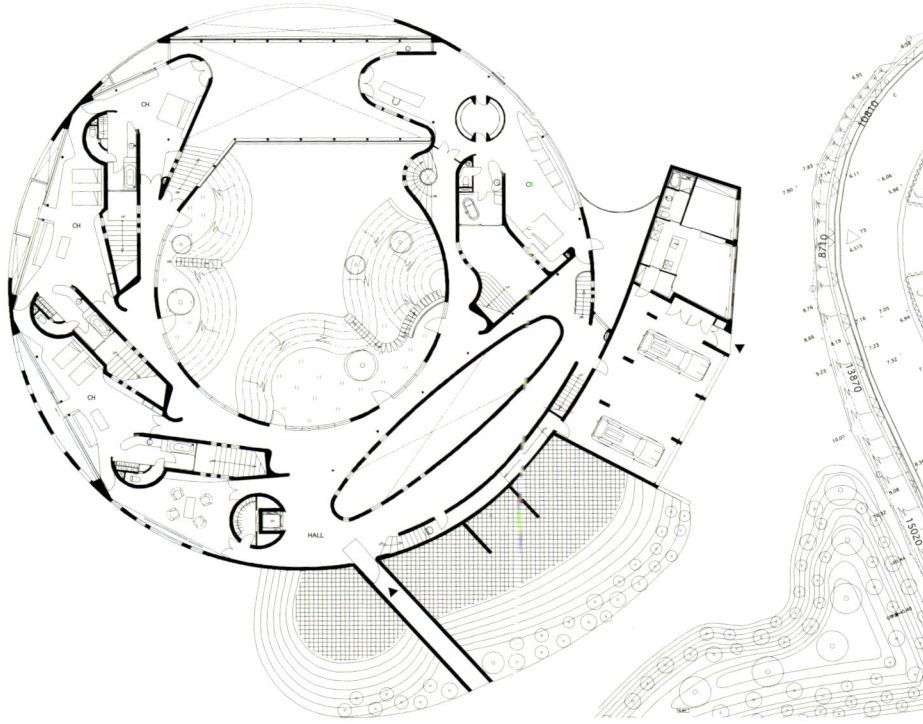

二层平面图 Second floor plan

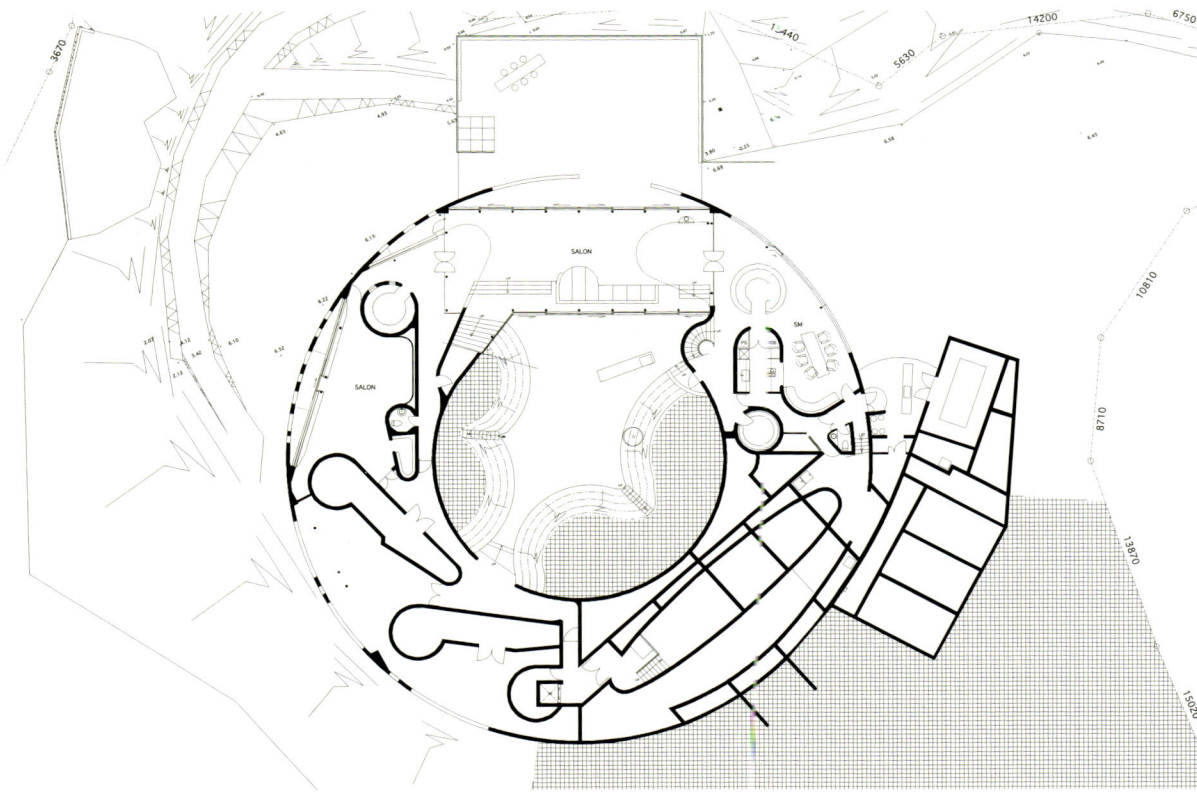

一层平面图 First floor plan

休息室 Lounge

餐厅厨房 Dining kitchen

浴室 Bathroom

书房 Study room

卧室 Bedroom

希尔顿芭提雅酒店
Hilton Pattaya

地点　Pattaya, Thailand
设计　Department of ARCHITECTURE Co.,Ltd. / Amata
　　　Luphaiboon, Twitee Vajrabhaya Teparkum
设计团队　Waraphan Watanakaroon, Prow Puttorngul,
　　　Tharadon Teerawanitchanan, Picha Thadaniti,
　　　Wipavee Kueasirikul, Sasicholwaree Sawatdisawanee,
　　　Rattanapon Monmahachinda, Sutah Schonrungroj,
　　　Atirojt Rojratanawalee, Worawut Oer-Areemitr,
甲方　Kanin Manthanachart
摄影师　Central Pattana Public Company Limited
　　　Wison Tungthunya

大厅和"移动"酒吧

大厅和酒吧位于酒店的 17 层，远远高于下面热闹的芭提雅海滩。人们从走廊的一端进入，走出电梯便来到了宽敞的大厅区域。整个天花板拥有动态的线条，指引游人走向远处的海岸。天花板上的布帘装饰是酒店的主要特色，而地面的简单布置营造了一派宁静的氛围。夜晚，上部长条状照明灯的光线透过布帘，使整个天花板散发出柔和的光亮，笼罩着整个大厅。在大厅的另一端，酒吧区沿着建筑的边缘而建，与海岸线平行，以便游客可以领略到最开阔的海景。酒吧的背景墙是木质的，带有壁龛，壁龛里有一排躺椅半嵌入墙体。宽敞舒适的家具为游客提供了舒适的座椅以放松身心。布置于酒吧一端的墙面镜使整个酒吧的视觉空间增加了一倍。室内酒吧的前方是一片户外休闲空间，里面有一个大型的水池，倒映着天空和躺椅，还有灯零散地漂在水面上。从这里开始，空间向海洋全景和轻拂的海风开放。

"边缘"餐厅

位于酒店 14 层的餐厅面朝大海，由多个开放式厨房组成，为游客提供世界各地的美食。餐厅的这种空间布局旨在最大程度地利用地理位置上的优势，使游客全方位地欣赏到迷人的海景。餐厅沿着玻璃窗设计，面向大海，层高 8m，就餐区占据了这个两层空间，餐厅后部的地板要高于前部，这种阶梯

式的平台设计让就餐者可以畅通无阻地观赏到室外美景。室外露台的基座略低于室内，以便游客可以一览无遗地欣赏到窗外的海景。位于 14 层的餐厅远离市井街道的喧嚣，为游客提供了安静、放松、舒适的环境。另外还有一个休闲空间，使用了自然材料和天然色调，使游客能安稳地放松、休憩。开放式厨房采用特殊材料布置，成为室内空间的中心。餐厅中还装饰着海扇和各种半透明发光的海洋生物，使游客仿佛置身于海底世界。内部表层几乎从原来的物质材料转化成柳膜包裹住整个空间。大小不同、颜色各异的几何形吊灯随意地悬挂在半空中，为整个空间提供照明。即便是在餐厅中的洗手间里，置身于这样的内部空间，也不禁令人想到海洋动物居住的海底世界。

Lobby & "Drift" Bar

The space for the hotel lobby and bar occupies the 17th floor, high above the bustle of Pattaya beach below. Upon entering the space from one end, as elevator doors open, one would enter a spacious lobby area. The architectural intervention to the entire ceiling plane, with its dynamic wave lines, leads the movement of the visitors towards the seafront beyond. The fabric installation on the ceiling becomes a main feature in the space while simple

elements on the ground provide a tranquil atmosphere. At night, strip lighting accents from above the fabric linear pattern. The whole ceiling volume becomes a gentle luminous source of light giving a fine ambience to the overall space. At the end of the lobby space, the bar area is arranged linearly along the building edge parallel to the sea with maximum opening to the ocean view. Backdrop of the bar area lies a wooden wall with alcoves where the daybeds partially tuck themselves into the wall. Oversized and soft furniture provides comfortable and relaxing seating for guests to sink into. A full-wall mirror at the end of the long space doubles the visual length of the bar area. Further in front of the indoor bar area is an outdoor lounge space with a large reflecting pond catching the reflection of both the sky and the droplet daybeds and lamps scattered around. From this area the space is opened up to the panoramic ocean vista and gentle sea breeze.

"Edge" Restaurant

A restaurant serving international food with multiple large open kitchens is situated cn floor fourteenth facing the ocean view. Its main spatial organization strategy is to open up the ocean view to the guests at its maximum to take advantage of the view from its prime location. The space is stretched linearly along the glass wall facing the sea with an almost 8 meter-high ceiling. The indoor seating area is organized into two tiers where the floor towards the back is higher up to ensure a good view over the front part. The outdoor terrace with its impressive panoramic view in the front lets the ocean vista flows uninterruptedly to the inside by arranging its floor plate a step lower. Away from the busy street down below, the restaurant provides a calm, relaxing and comfortable atmosphere. An airy space with the use of natural materials and light colors allow the guests to set back and relax. The open kitchen area is accentuated with special material treatment as a focal point of the interior space. The visual elements in the space are loose reminiscent of an underwater landscape – sea fan and translucent luminous ocean creatures. The interior surfaces are almost transformed from their original materiality into thin gorgonian membranes wrapping the space. Clusters of glowing organic-shape lamps suspended randomly in mid-air with varying sizes and colors scatter throughout the space. Hidden in the restaurant restroom, maneuvering through its interior space, one cannot resist thinking of the space in between the seabed fauna.

户外休息室（夜景）2 Outdoor lounge(night) II

户外座位区 Outdoor seating area

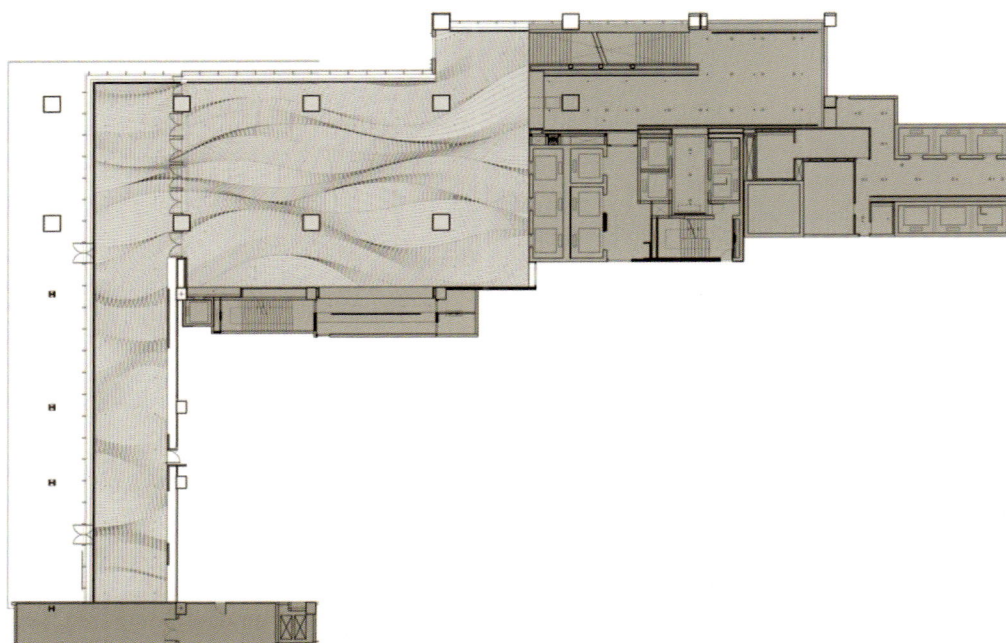

天花板平面图（大厅）Ceiling plan(lobby)

大厅 Lobby

电梯厅 Elevator hall

餐厅连接处 1 Linkage to restaurant 1

餐厅 2 Restaurant Ⅱ

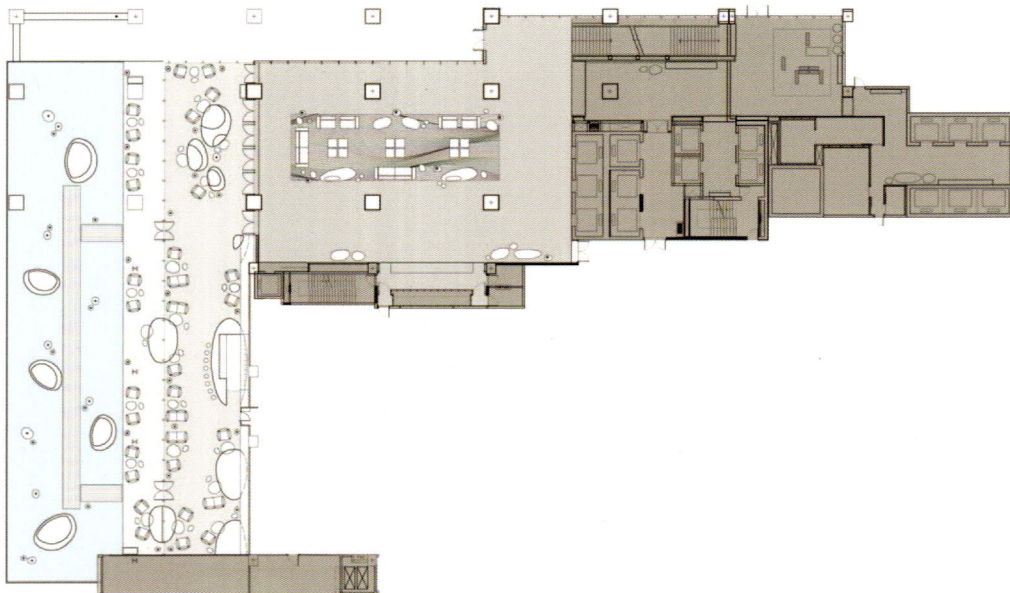

楼层平面图（大厅） Floor plan(lobby)

餐厅3 Restaurant 3

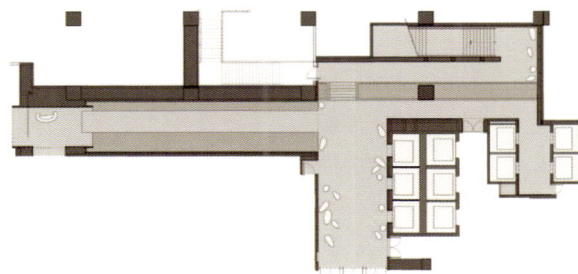

楼层平面图（电梯厅） Floor plan(elevator hall)

0 5 10(M)

餐厅 4 Restaurant IV

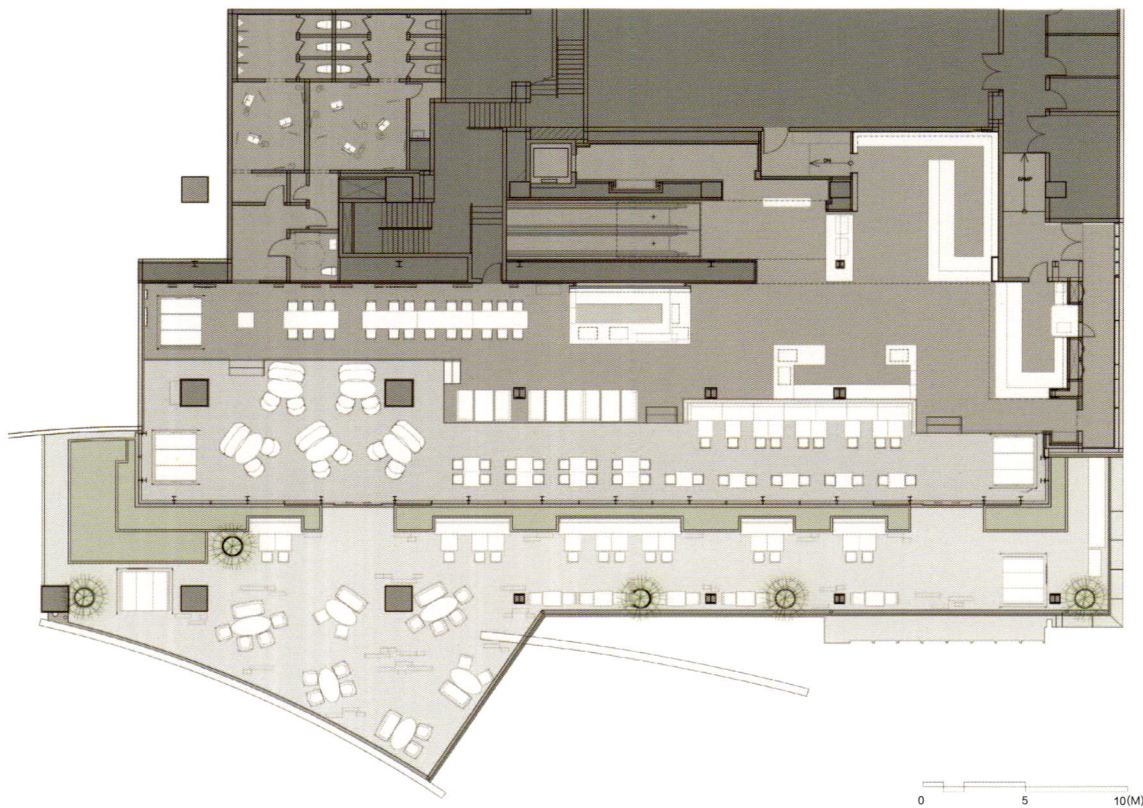

0   5   10(M)

楼层平面图（边缘）Floor plan(edge)

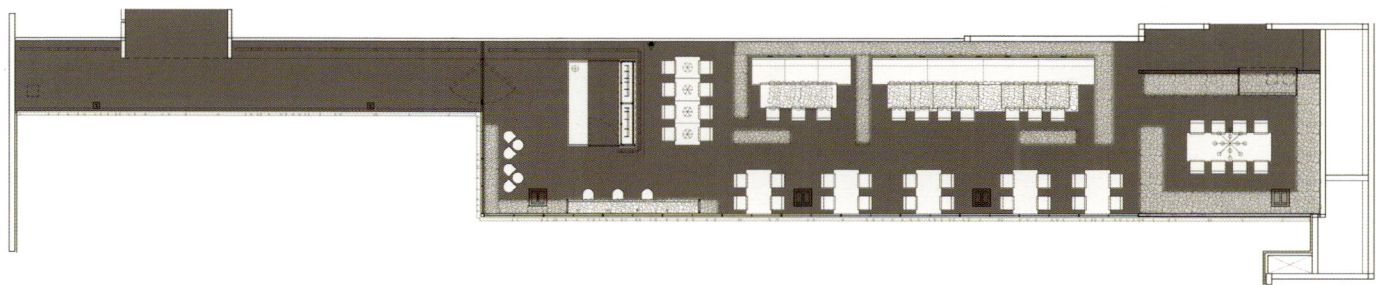

楼层平面图　Floor plan(flare)

开放式厨房　Open kitchen

玛雅里维埃拉东方酒店
Mandarin Oriental
Riviera Maya

地点　Playa Del Carmen, Quintana Roo, Mexico
室内设计　Henriksen Design Associates, Inc., Baia Arquitectura
建筑师　Desarrollos Marinos del Caribe (DEMA) + Gilberto Borja
摄影师　Mandarin Orienta

总统别墅 Prasidential villa

大厅 Lobby

艺术庭院露台 Art courtyard patio

玛雅里维埃拉东方酒店坐落于自然水域与幽静自然的完美结合处，酒店的 128 间豪华定制客房坐落于现代化的一层及两层高的别墅中，为游客提供了极其宁静祥和的环境。从酒店所有的客房都可以看到极具魅力的水景、宁静的泻湖、蜿蜒的水道或是加勒比海。酒店的内部装修以当地颇具异国情调的树木与岩石、硬竹木地板和奶油色大理石为原料，还运用了精湛的墨西哥艺术，体现了亚洲特色与热带风情的完美融合。

作为拉丁美洲真正独一无二的旅游度假胜地，玛雅里维埃拉东方酒店中占地 2322.6m² 的水疗中心全面设置了健康新标准。水疗中心由包含 3 个贵宾套房在内的 11 个理疗室、占地 139.4m² 的墨西哥最先进的健身中心、占地 92.9m² 的瑜伽练习室和一个充满阳光的室外温水游泳池组成。

酒店内的各式餐厅和酒吧不仅为游客提供各式精美菜肴，还有宜人海景、室外雅座与动态设计，在这样的环境中用餐，游客即是置身于旅游胜地的创新性与想象力的最前沿。Ambar 餐厅以铜质屋顶、宽敞的酒窖和开放式厨房为主打。它的正上方就是别致的中式酒吧，一个功能多样的户外酒吧和一个酒廊为游客提供各式极具异国风情的饮料。海洋全景造就了沙滩餐厅 Aquamarina，它拥有海蓝色与银灰色为基调配合实木装修的超现代简洁风格，使游客在户内外就餐都成为享受。

Perfectly placed along natural waterways that blend seamlessly into the secluded setting, the 128 luxuriously appointed guest rooms are situated in contemporary-styled bungalow villas or two-storey villas, offering the ultimate in peace and tranquility. All guest rooms enjoy exceptional water views of the mesmerizing cenote, the tranquil lagoon, the meandering waterways or the Caribbean Sea. The interiors are exquisitely designed with Asian influences and tropical flair, using native exotic woods and stone, bamboo hardwood floors, crème-colored marble and an exquisite showcase of Mexican art.

Truly unique to any destination resort in all of Latin America, the 25,000-square-foot Spa at Mandarin Oriental Riviera Maya sets new standards of holistic well-being. The spa sanctuary features 11 treatment rooms including three VIP couple suites, a 1,500-square-foot state-of-the-art fitness center, a 1,000-square-foot yoga room, and a heated outdoor pool drenched in natural sunlight.

With dining at the forefront of the resort's innovation and imagination, a variety of restaurants and bars serve an eclectic selection of contemporary cuisine in settings inspired by outstanding ocean views, outdoor seating and dynamic design. Ambar's unique design boasts a copper roof, walk-in wine cellar and show kitchen. And, just above Ambar is the very chic Mandarin Bar, a sophisticated outdoor bar and lounge serving exotic drinks and cool ocean breezes. Panoramic ocean views define the beach club restaurant Aguamarina, with its striking, ultra-modern, minimalist design colored with ocean blue, sleek silver and natural woods, and both indoor and al fresco dining.

高级灰岩房间 Premier cenote room

高级灰岩浴室 Premier cenote bathroom

热带雨林房间 Selva room

滨海小屋 Beachfront casita

总统别墅——起居室 Presidential villa - living room

双人桑拿套房 Spa couple suite

桑拿套房——休闲区 Spa suite – relaxation area

桑拿——活力池 Spa – vitality pool

Aquamarina 餐厅 Aquamarina

伦敦 W 酒店
W Hotel London

地点　10 Wardour Street, London, UK
设计　Concrete Architectural Associates
设计团队　Rob Wagemans, Jeroen Vester, Ulrike Lehner, Erik van Dillen, Melanie Knüwer, Jari van Lieshout,
Sonja Wirl, Nina Schweitzer
建筑面积　8,100m²
饰面材料　Granite Tile, Rubber Flooring, Timber Flooring, Wood, Tile, Stone Tile, Carpet
Wall : Glass, Mirror, Panel, Aluminium, Tile, Laminate, Timber, Plastic, Plaster, Paint, Carpet, Leather,
Curtain, Wallpaper
Ceiling : Plaster, Paint, MDF, Timber, Mirror Tile
施工单位　Jestico+Whiles
摄影师　Ewout Huibers

休息室剖面 Section – Lounge

游客走过由 280 个小闪光球拼接而成的闪光球云雕塑便可进入酒店。酒店的墙壁由黑色玻璃建造，配合从闪光球云里射出的光点，使酒店入口光芒闪烁。踏入伦敦 W 酒店就仿佛步入了一个崭新的世界。沿着闪光球云雕塑向上走便来到了酒店的一层。酒店迎宾区有三个服务区，每个服务区的组建模块相同，但组合方法不同，因而产生了三个形态各异的区域；从顶棚与地面不同角度所发出的柔和的紫色光芒充溢着这三个不同的区域。

宽敞的走廊从迎宾区延伸到酒店的另一侧。一个巨大的白色英国国旗式橱柜将第一部分隔开，你可以在 W 休息室中坐下来，随手翻阅书籍。乍看来，W 休息室的座位和灯光仿佛是一体的，但细细看来你便会发现，休息室中的每一处都别有洞天。每一处雅座都有着独具一格的装饰风格和与座位形状一致的发光体。位于同一时空却有着不同设计风格的对比最大限度地迎合了客人们的需求。W 休息室的酒吧并不是一个真正意义上的酒吧，37m 长的切斯特菲尔德沙发是游客与朋友们聚会、交流、谈天的绝佳去处。酒吧中包括橡木地板和金箔天花板在内的所有装饰都自然地与曲线形景观配合。这是一个在任何时间都可以享受生活的绝佳地点。酒吧可以在不同位置装上垂帘，随着一天之中时间与空间用途的变化创造出不同的光线效果。

在伦敦 W 酒店的 WYLD 酒吧中，客人们可以劲歌热舞直至天明。WYLD 酒吧配有红地毯、红黑相间的皮质沙发、直径达 3m 的迪斯科闪光球。一个圆形吧台笼罩在红色灯光中为客人提供各式定制鸡尾酒。酒吧四周的墙壁上镶嵌着黑色亮片，象征着伦敦西区，并随着音乐节奏的变换而闪烁。

客人还可以在酒店的 Sweat 健身中心和 AWAY 温泉中心享受到高品质的服务。Sweat 健身中心内宽敞明亮，俯瞰着整个莱斯特广场。Sweat 健身中心内铺设着深灰色的橡胶地板及地毯，既可以降低噪音又为伸展、跳绳等运动增加舒适度。游客还可以使用健身房中的各种健身器材锻炼。Sweat 健身中心内闪烁的光线受到了闪光球与 WYLD 酒吧中异彩纷呈的迪斯科球的启发。需要轻微放松的游客可以去往 AWAY 温泉中心。穿过银白色的帘幕便是温泉中心的接待处，在这里你可以预定推拿按摩或进入桑拿浴室和蒸汽浴室。

伦敦莱斯特广场上的 W 酒店的客房既是舒适的休息处又极尽奢华。进入客房，首先呈现在眼前的是化妆室。化妆室的正中摆放着白色的奢华桌足，以便于客人整理仪表。巨大奢华的桌子足以摆放客人的首饰、化妆用品及笔记本、电话、文件等各种办公用品。卧室内一系列用品均设计得舒适松软，为客人提供了安心舒适的环境。卧室地面铺有松软的棕色长毛地毯，以隔绝地板的凉气；墙面悬挂着大幅帷幔，以遮盖住冰冷的墙壁。卧室内部舒适简洁的设计与化妆区内闪亮的镜面墙壁形成了鲜明的对比。

这套顶级客房带有大厅、宽敞的客厅、独立餐厅、主卧，还有自己的化妆天地。WOW 的顶级套房并不是专为名流而设。入口大厅连接着位于客房中央的客厅、餐厅与卧室。大厅的墙壁为香槟色人造皮革拼接装潢，还设有一面独立的巨大镜面墙壁。客厅中央摆放着圆形旋转式的切斯特菲尔德沙发，沙发下垫着柔软的黑色地毯，沙发上方悬挂着光芒闪烁的球形雕塑。客房可以在不同位置装上垂帘，随着一天之中时间与空间用途的变化创造出不同的光线效果。

Guests make their entrance underneath a disco ball cloud sculpture made of 280 disco balls. Supported by black glass walls and dynamic spots that point towards the cloud, this hotel entry becomes a dazzling room of reflections. To walk inside this hotel is like stepping into a new world. The disco ball-cloud sculpture guides you up the first floor. The welcome area holds three circular pods for guest services. All pods have equal modular parts, but each has been stacked differently to create three unique shapes. A soft purple light supports the different shapes and shines from underneath and above each pod.

A broad passage stretches out from the welcome area to the other side of the hotel. A grand white Union Jack cabinet separates the first section where you can pick a book to read and take a seat in the archipelago of W Lounge. The seats and lights of W Lounge appear as one common piece of land at first sight, but once you get closer they all turn out to be different hideways. Each island has its own furnishing design and a mobile-like light object that follows the shape of the seating. This contrast of being in the same space and time but on a differently designed level, tries to comfort guests in either way.

W Lounge bar is not so much about the bar; it is the 37m-chesterfield couch that defines the social landscape for guests and friends to live, meet, mingle and flirt on. Everything, including the end-grain oak flooring and gold leaf ceiling, naturally follows the winding landscape. The lounge bar is a good place to stay at any hour of the day. A striking framework of vertical blinds can be set in different positions and creates different light scenes according to the time of the day and use of the space.

Guests can dance the night away at WYLD bar. WYLD's interior look combines the red carpet feel with spicy red & black leather furnishings and a grand finale 3m diameter disco ball. A circular booth surrounds bespoke cocktail tables that light up in red. The high level walls are covered with black sequins referring to the West End and move according to the beat of the music.

Guests can charge up at the hotel's Sweat and AWAY Spa. The Sweat is a very light and spacious gym overlooking Leicester square. It has a dark grey rubber flooring that deadens sounds and cushions steps and is comfortable to exercise, stretch or skip rope on. Guests can use Technogym sporting equipment for a good work-out. The sparkles of the light are inspired by the disco ball cloud and extravagant WYLD bar disco ball. Guests who need a little extra relaxation can enjoy it at the AWAY Spa. Curtains of white and silver strings make a delicate entrance to the reception desk, where you can book a massage or a visit to the sauna or steam bath.

W London Leicester Square guest rooms combine a comfortable sleeping zone with a luxurious vanity area. When you enter the room, you step into the dressing area. A white monumental vanity desk with matching vanity chair has been placed central of the area as the supporting piece for dressing up activities. It provides more than enough space for displaying fashion accessories and make-up articles as well as notebooks, phones and some paperwork among other things. Bedroom area offers soft surroundings to make you feel at ease and ready to relax. Here you have a lush, high pile, brown carpet to warm feet and large curtains to cover up the normally cold walls. The bedroom interior is a soft antidote to the shiny mirror walls of the dressing room area.

An extreme suite with entry lobby, gargantuan living room, private dining, master bedroom and last but not least your own spectacular dressing heaven. The Extreme WOW suite is not for wallies. The entry lobby connects to the central living room, dining area and bedroom. Its walls have been upholstered with champagne artificial leather stitching and there is a separate giant mirror wall. The living room center-piece is a splashing circular and rotating chesterfield couch with lush black carpet underneath and glitter ball sculpture hanging above. A striking framework of vertical blinds can be set in different positions and creates different light scenes according to the time of the day and use of the space.

大厅 Lobby

休息室 1 Lounge I

休息室 2 Lounge II

休闲美酒吧 Lounge bar

香料市场 Spice market

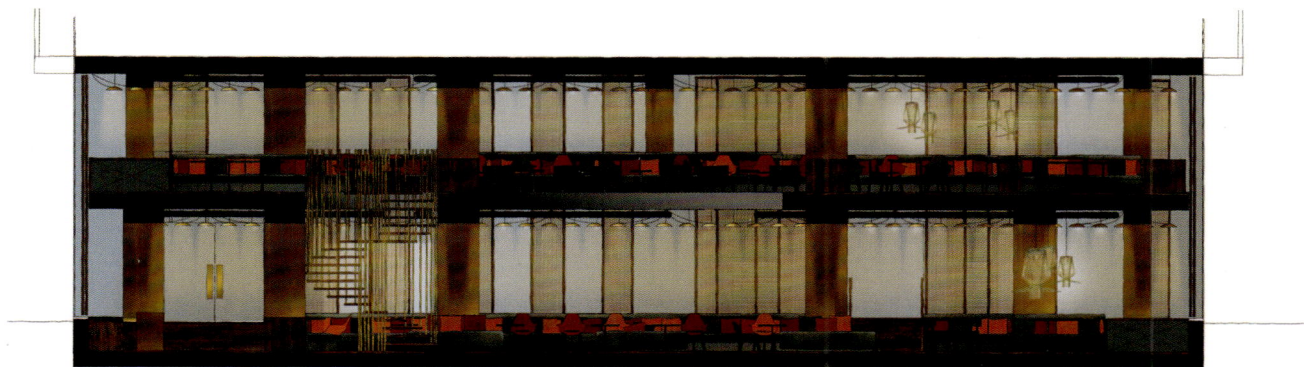

市场剖面 Section – Market

EWOW 客房 Ewow suite

桑拿套房 Spa suite

豪华套房 Wonderful room

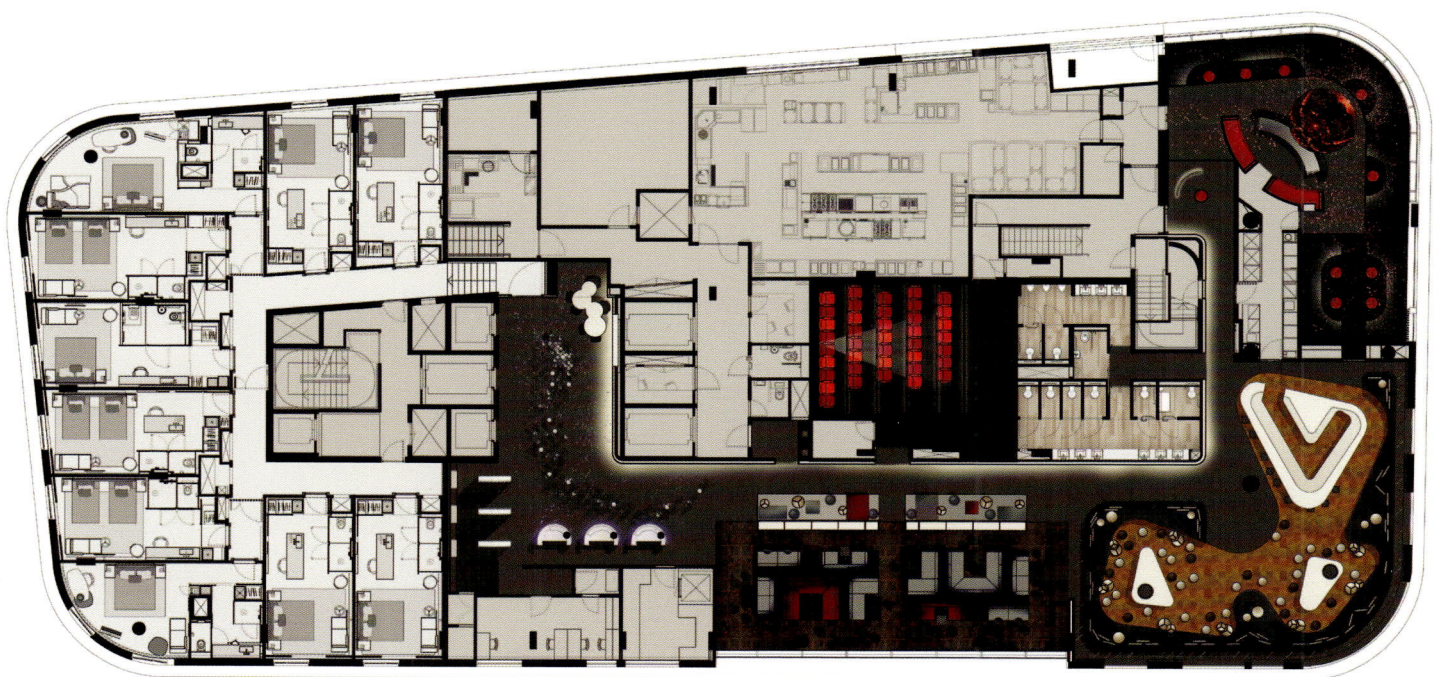

二层平面图 Second floor plan

nhow 酒店
nhow Hotel

地点　Berlin, Germany

设计　Karim Rashid Inc. / Karim Rashid

饰面材料　Laminate Wood Flooring, Carpet, Tile, Wall Covering, Barrisol, Paint, Plastic Laminate, Curtain(int.)

参与设计人员　Project manager_Camila Tariki, Architect_Kamala Hutauruk, Cece Stelljes, Julie Lee, Evan McCullough

施工单位　BAM Deutschland AG

摄影师　Offered by Karim Rashid Inc.

外观 Exterior

柏林 nhow 酒店由 Karim Rashid 事务所设计，它不仅仅体现了柏林现代的时代精神，还创造了数字艺术与空间组成的科技有机的建筑，从而与世界其他的地方产生了联系。
客人们在引人注目的接待处会受到热烈的欢迎。具有感性美的雕塑由高光玻璃制成，里面还安装了照明设施。朦胧的颜色和柔和的曲线使得 nhow 酒店的休息室内充满了感性的氛围。在 Karim 设计的家具中依然具有这种柔和的风格。天花板看上去好像是融化状态的塑料，极具雕塑感。镶嵌在里边的灯使它看上去像流动的一样。大堂一直延伸到 nhow 酒店的酒吧。黄金漆玻璃纤维制成的无所不在的天花板使得酒吧看起来很豪华。在酒吧的侧面是座位区，由奢华的符合人体结构的沙发和躺椅组成。定制的波浪状凳子与雕塑感的天花板相呼应。完全定制的带有 Digipop 图案的窗帘作为一种艺术品，为施普雷河的美丽景色增加了更多的色彩。
nhow 酒店餐厅的灯光选择了柔和的色调，因为墨绿色有助于消化，淡粉色有助于舒缓情绪。有机形状的灯具照亮了整个餐厅。在餐厅的中间有一些诱人的雕塑，它们既是艺术品，又是使用物品，集形式和功能于一身。这些雕塑由涂漆的玻璃及玻璃纤维制成，人们可以在上面享用早餐和午餐，晚上空闲不用时就是一件可以欣赏的艺术品。因为餐厅有一个推拉式的隔墙，因此可以在这里举行私人活动。旅行者和当地人可以坐在公共的桌子旁轻松就餐。还有更小的私人桌子为人们提供一种更私密的就餐空间。柏林的 nhow 酒店有六个会议室，每个会议室内都有 LED 屏滚动发布会议和活动信息。
图案极具活力的壁纸和地毯指引着客人穿过走廊到达客房。每一个楼层的电梯都有不同的 Digipop 图案，到处充满着愉快的气氛。一个漂亮的酒店不仅关注人们的需求，还关注人们的意愿和行为。酒店就是为了舒适和充满能量而设计的。东楼的房间使用了太阳升起和降落时温暖的颜色——粉红色和金色的混合色。西楼的房间由比较冷的颜色装饰——灰色、蓝色和粉色混搭。在酒店的最高层，最上部的房间可以看到最好的施普雷河景色，并且都用比较冷的颜色装饰，特意为商业人士和偶尔"开溜"的人准备。每间客房都把一些大胆的 Digipop 图案抽象艺术品作为床头，这能激发人们做个好梦。房间里定制图案的地板是由那些被粉刷过的可以再利用的层压木板制成的。Karim 设计了一种双面镜，当电视不用时可以把电视屏幕隐藏起来。镜子好像是附在了一个凸出于墙外的桌子上。浴室的墙壁是白色的倾斜玻璃，所有的细节设计都非常引人注目，让自然光进入的同时还能保护客人隐私。照明设备可以进行个性化调节，制造出不同的心情：忙碌、高兴和轻松。
在高档的房间里，曲面隔墙把卧室和起居室分隔开。镶嵌在墙壁里的电视机可以旋转，可以在床上看，可以在隔出来的起居室的真皮沙发上看，也可以在具有雕塑感的纤维玻璃桌子上边工作边看。宽敞的浴室里有一个定制颜色的独立浴缸，还有很大的玻璃淋浴间、马赛克瓷砖地面、两个水槽和奢华的 nhow 设施，人们可以在这里尽情享受。

nhow Berlin designed by Karim Rashid celebrates Berlin's modern zeitgeist and connects it to the rest of the world by creating a technorganic land of data-driven art and spaces.
Guests are greeted by the visually arresting reception desk. The sensuous sculpture is made of high gloss fiberglass with inset lighting. Twilight colors and soft curves create a sensual mood in the nhow hotel lounge. The soft language continues into the Karim designed furnishings. The ceiling features a sculptural installation of almost molten plastic. Recessed lights give it a dynamic glow. The lobby flows into the nhow hotel bar. The omnipresent and omnidirectional head

made of gold lacquered fiberglass tops the bar creating a lavish statement. Seating areas composed of the voluptuous organic and ergonomic couches and lounge chairs flank the bar. Custom wave banquettes mirror the ceiling sculpture above. Sheer custom curtains with a Digipop pattern act as artwork coloring the views of the iconic view of Spree River.

The light filled nhow hotel restaurant's pastel color palette was chosen because green shades aid digestion and light pink is soothing. Organic shaped lighting fixtures illuminate the room. The striking sculptures in the middle of the dining room serve as both Art & Object, Form & Function. Made of lacquered fiberglass and glass they serve breakfast, lunch and when not in use at night, simply be an art element. A sliding partition makes the space reconfigurable for private events. Communal tables bring together travelers and locals in a casual setting. Smaller private tables offer a more intimate dining experience. The nhow hotel Berlin boasts 6 conference rooms, each with scrolling led signage announcing meeting and events. Energetic patterns in the form of wallpaper and carpets carry guests through the hallways to the guests' rooms. Each floor features a different custom Digipop patterned elevator bank making every moment entertaining. A beautiful hotel embraces human needs, desires and behaviors. Designed for comfort and positive energy. The East Tower rooms feature a sunrise/sunset color scheme of warm gold and invigorating pink. The West Tower Technolux rooms have cooler color schemes of grey, blue and pink. High aloft the hotel, the Upper towers have premium views of the Spree River, and are awash in calm colors, ideal for business traveler or peaceful get-away.

In each guest room bold Digipop artwork serves as the bed headboard, imagery meant to inspire dreams. Underfoot the custom patterned flooring is made of printed recyclable wood laminate. Karim designed a two way mirror to hide the flat screen television when not in use. Remarkably the mirror seems to morph into the chrome desk cantilevered from the wall. Leaving no detail unnoticed, the bathroom walls are white gradient glass allowing natural light in while still giving guests privacy. The lighting can be personalized to create different preset moods: work, play and relax.

On premium rooms, a curved wall divider separates the bedroom from living area. An inset television rotates within the wall for viewing in bed or while lounging on the leather couch in the separate living room or working from the sculptural fiberglass desk. The spacious bathroom features a freestanding bathtub in custom colors, large glass showers, mosaic tile floors, double sinks and the luxury nhow amenities to pamper one's self.

接待处 Reception

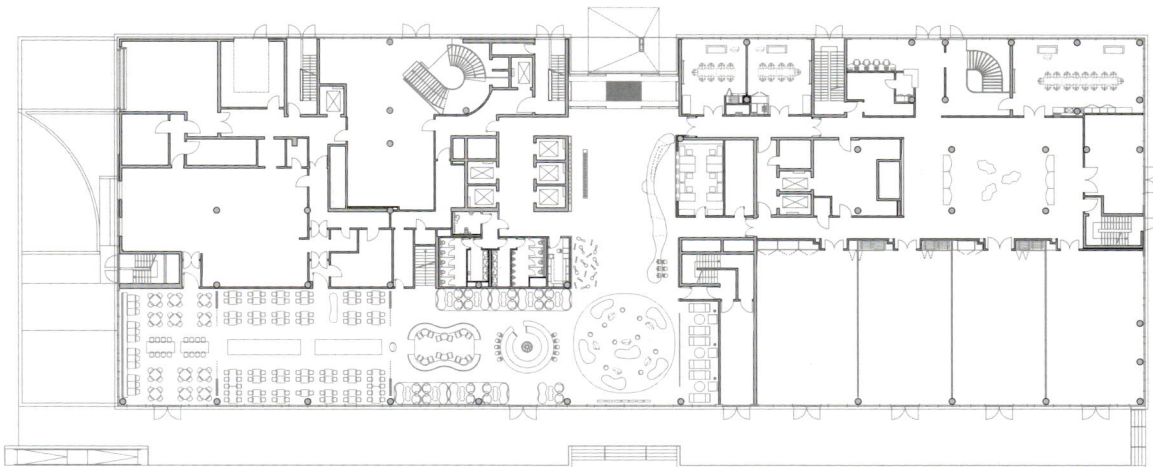

一层平面图 First floor plan

吧 Bar

餐厅 1 Restaurant I

餐厅 2 Restaurant II

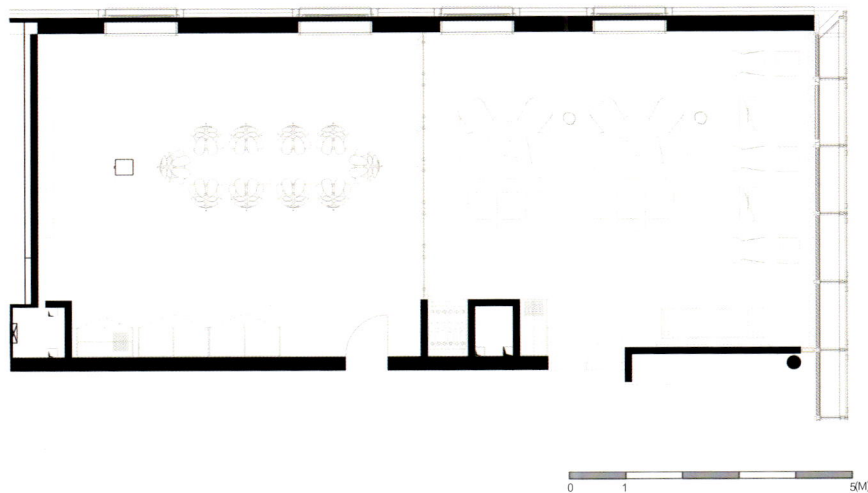

楼层平面图（休息室） Floor plan(Lounge)

套房1 Suite I

套房2 Suite II

起居室 Living room

卧室 1 Bedroom I

卧室 2 Bedroom II

1 起居室 / 休息室

2 餐厅

3 餐具室

4 储存室

5 卫生间

6 起居室

7 卧室

8 步入式衣橱

9 浴室

1 Living room / Lounge

2 Dining Room

3 Pantry

4 Closet Area

5 Restroom

6 Living Room

7 Bedroom

8 Walk-In Closet

9 Bathroom

楼层平面图 1 Floor plan I

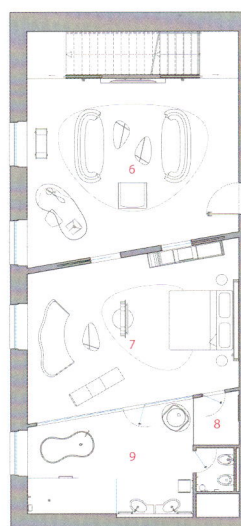

标准间/Standard room

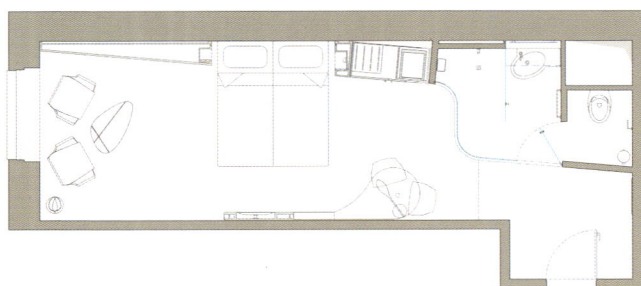

0   1   5(M)

楼层平面图 2 Floor plan II

淋浴室 Shower

浴室 Bath room

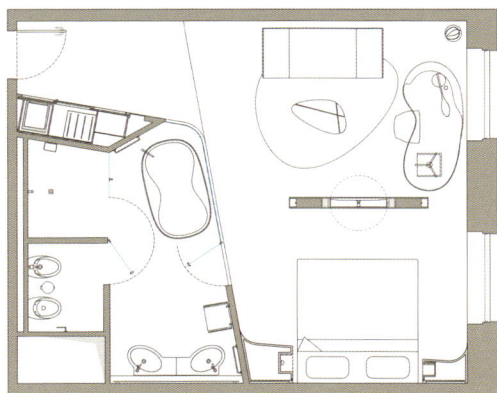

楼层平面图 3 Floor plan III

楼层平面图 4 Floor plan IV

Art Del Teatre 酒店
Hotel Arc Del Teatre

设计 EQUIP · Xavier Claramunt, Martin Ezquerro, Pau Rodríguez, Marc Zaballa
地点 Barcelona, Spain
总楼面面积 4,744m^2
设计团队 Yago Haro, Miquel de Mas, Joan Cuevas, Javier Luri, Ho-Sang, Anne-Sophie
de Vargas, Oliver Schmidt,
Carmen Barberà, Vicky Pons, Adam Mendoza, Sandra Yubero, Alex Ortiz,
Hilda Compte, Oriol Bordes, Anna Ramos
摄影师 Adrià Goula

城市依旧在给我们呈现不同样子。现阶段资源节约正被大力倡导着，我们建设出极不寻常并且极具表现力的建筑。

回顾历史，有一条街道邻近 Parallel 街，位于迷宫一样的"Raval"街道网格中，被称为"Barrio Chino"，那里曾经坐落着一个剧院。这所建筑曾经被认为是在巴塞罗那市建造一个全新旅店的机会。

关于这个建筑的主题，设计最开始采用了两种典型的战略：倾听这个地方，询问使用者意见。为了更好地解释这种概论，我们可以解释为一个人用耳朵倾听这个地方，对两墙之间的建筑进行讨论。换句话说，这个方案保留一种历史外观，尽管它不是由官方来保护，而且还将建筑插在其与相邻建筑的共用墙之间。新建筑通过对房间外观的处理，使得整体上看上去好像有很多像井口一样格子的书架一样，这采用了和旧时期建筑完全一样的风格，具有很高的聚集度，使得那些位于街道水平之下的公共区域给人留下戏剧般的环境氛围。整齐、静态的外观是一种传统式房间的方案，它和这个区域其他建筑的外观的关系很复杂，而且充满变化，这种整齐、静态的外观使用一种巨大的遮盖，可以反映出新建筑物外表光线的变化。

这些方案都是关于空间的，但是也有关于整合方面的。不乐观且直接地说，我们不得不承认保护这种外观是将它作为曾经在这里存在建筑物的一种纪念，同时也把它当做是和法规及安全权限交涉的工具，这样可以隐藏一些建筑物，从而形成一个门廊，在这个迷宫般的街区内建立了一个内部平台。和法规的交涉，使得我们可以在底层建立一些房间，而这些房间拥有特殊聚类系统，使得这些房间具有良好的私密环境，光线可以完全照到各层房间内。

使用这种巨大的遮盖盖住建筑物的外表的目的是当屋内和屋外的光线变化时，无论是从房间里面看还是从街道上看，建筑物的外观都在改变。这将周围建筑的多样性都集中在了一起。

在这个建筑物建造过程中提倡的资源的节制和节约理念在这个巨大的遮盖物的使用中也得到体现，它将随着季节的变化而变换，可以区分自然光和人造光，并选择性透过。

这座新的旅馆有两种不同的外表，这些外表取决于使用者是在房间内还是在公共区域的任何一个地方。建筑物内的地下公共场所，被分别装修成为舒适的角落和不同环境。通过使用各种形状和各种颜色的家具来增加一种活力，尽管在现实视野中留意不到。相反，房间充满着冷静的气氛，这样有助于旅游者调整决定。通过建筑侧面区的淋浴区，可以直接建立房间内部和街道的联系，用塑料制成的巨大遮盖物使得街上的人不会看到房间内的一切。

旅馆尽管作为一个给旅游者提供休息的地方，但是旅馆也在试图保留城市特点，通过公共区域内的物体和组合以及从房间内可以看到街道的视野来保持室内和室外的联系，因此这个建筑物优点类似进出街道的门，同时形成了一种私密并且另类的历史区域。

The city still puts on quite a show. Struggling insolently with the prevailing imperative of sobriety and economy of resources, we are presenting an outrageous and exhibitionist building.

Looking back over history, it turns out that in a street very close to Parallel, but well within the maze of streets known as the Raval, and particularly known as "Barrio Chino", there once stood a lofty theater. This building has been seen as the opportunity to build a new hotel that arrive in the city of Barcelona.

And on the subject of the building, two classic strategies are initially applied: listening to the place and asking the user. To further define this generality, we would say that one listens to the place with one's ears and talks by building between two walls. In other words, the decision was taken to preserve a historic facade, despite the fact that it is not officially protected, and to insert the building between it and the party wall of the adjacent building. The new building treats the rooms as if they were shelves, using wells, a resource borrowed from the selfsame historic buildings of the highly compact old district to afford a dramatic ambience to the common spaces, which are located below street level. The relationship of the neat and static facade which would yield, and does yield, a programme of conventional rooms, with the other facades in the area is imbued with a certain complexity and change, using a huge wrap that reflects the changing light of day on the facade of the new building.

These decisions are all about space, but they are also about integration. Being very sombre and direct, we would have to say that the facade is conserved as a memory of the building that once stood there, but also as a tool for negotiating with the regulations and securing permission, yes sir right away yes sir, making it possible to recess the building on the ground floor, form a porch and open up an inner patio inside the utterly dense maze of streets in the area. Working thus with the legislation will render it possible to locate rooms on the ground floor, in turn solving the conditions of privacy with peculiar clustering systems, with light entering radically and split-level floors.

The aim of the huge wrap covering the facade is for the building's appearance to change during the day both from inside the rooms and from the street while lighting conditions also vary on the inside and on the outside. It is a commitment to integration with the variety of the surrounding buildings. The sobriety and seriation called for by economy of resources applied in the construction of this building is granted a degree of variety by this huge wrap, which will change with the seasons and will filter the natural light of day differently to the artificial night-time light.

Asking the user, the new hotel has two different miens depending on whether you are in the room or in any one of the common areas. The public areas in the building, below street level, are treated heterogeneously to generate cosy corners and different ambiences. A variety of forms, colors and furniture is offered, as an extension of the street-level dynamism that is not actually within sight. And on the contrary, the rooms are imbued with sobriety to make for easy adaptation to the tourist's decisions. From the inside of the room a direct visual relationship is established with the street through the shower area, located on the facade, and the huge wrap which, with the help of vinyl, protects it from the eyes of the street.

The hotel, albeit a place of rest for the tourist, seeks to remain ultra urban, peppering the inside with a dose of the outside, with objects and combinations in the public areas and views of the street from the rooms, hence the building is akin to a door onto the street and, in turn, onto the hidden, private and heterogeneous parts of a historic district.

立面 Facade

外部（夜景）External Facade (night)

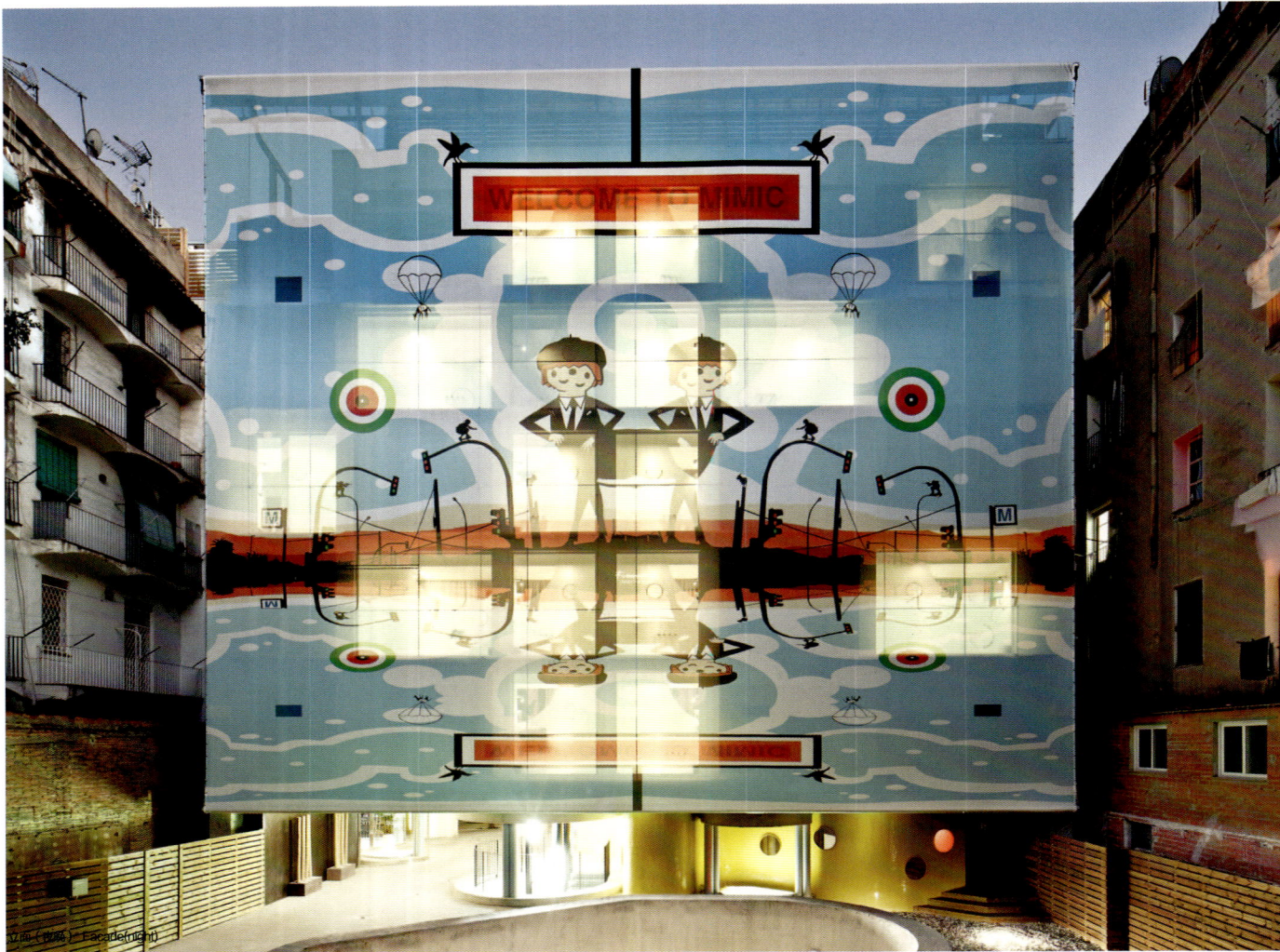

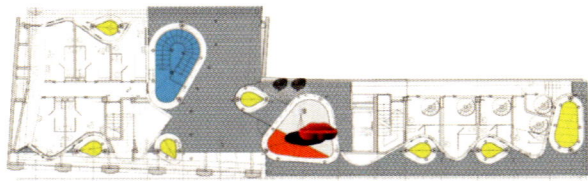

图像映射图 Image map

一层概貌 General view of 1F

一层接待区 Reception in 1F

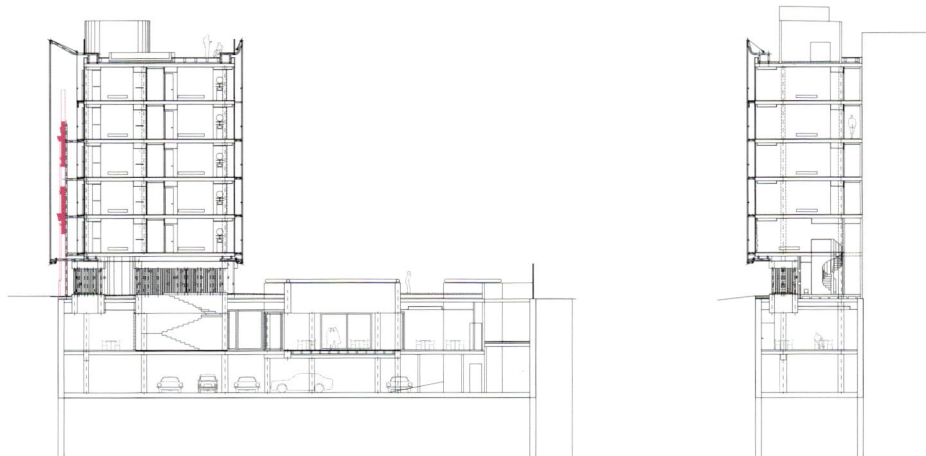

剖面图 Section

地下一层休息室 Lounge in B1F

立面图 Elevation

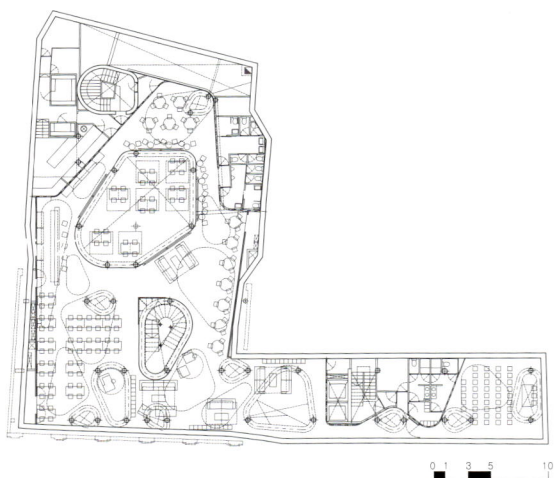

地下一层平面图 Basement first floor plan

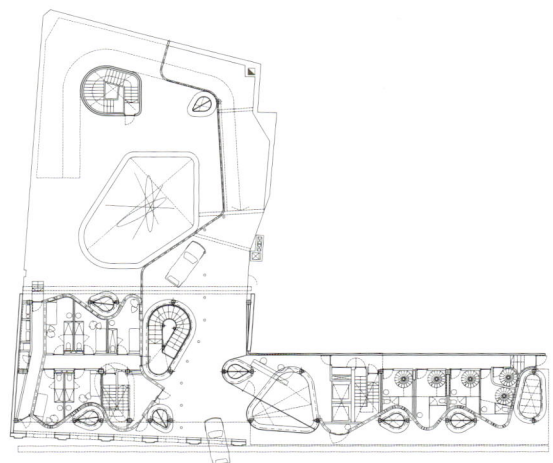

一层平面图 First Floor plan

卧室 1 Bedroom I

三层平面图 Third floor plan

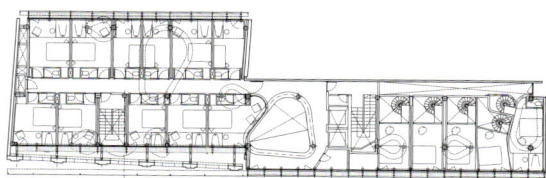

二层平面图 Second floor plan

卧室 2 Bedroom II

Bayside Marine 酒店
Bayside Marina Hotel

地点	Kanagawa, Japan
设计	Yasutaka Yoshimura Architects / Yasutaka Yoshimura
功能	Accommodation facility
基地面积	7,426m^2
建筑面积	1,720m^2
总楼面面积	2,087m^2
结构	Steel
楼层	2FL
景观设计	Studio Urban House Inc., GLAC
结构工程师	Jun Sato Structural Engineer
建筑设备	Kankyo Engineering
总承包商	Hazama Corporation
单元建造地点	Soleil(Thailand)
摄影师	Yasutaka Yoshimura

N

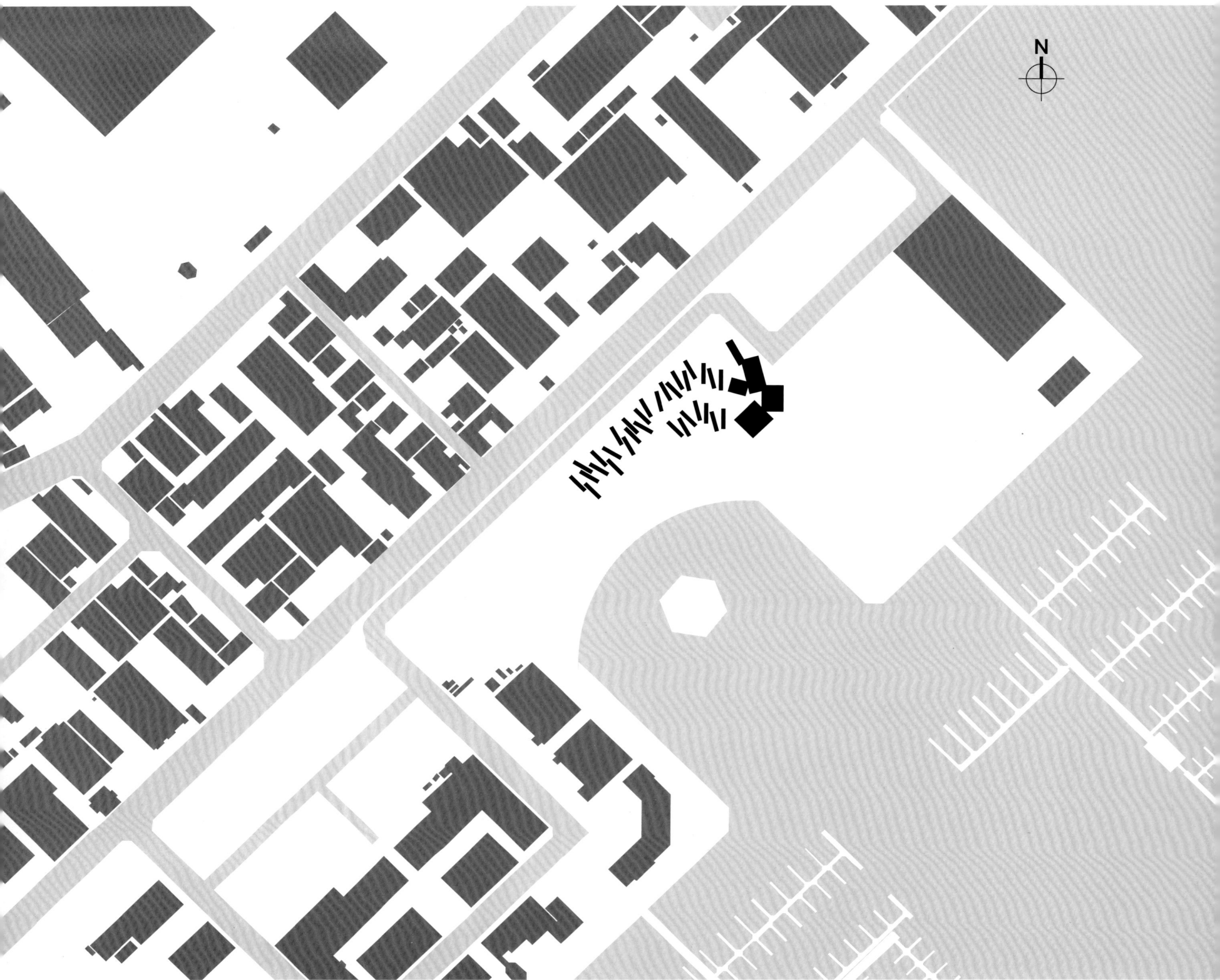

总平面图 Site plan

外观1 Exterior 1

外观 2 Exterior II

集装箱大小的别墅

横滨的 Bayside Marina 酒店是一项海滨别墅旅店工程。这项工程使用一种标准的 40ft（大约 2.4m×12m）的货船集装箱作为它的构件单元。一个构件单元提供了一个公寓式房间，两个组合的单元提供别墅式的房间，总共有 31 个房间。这些构件单元计划使用集装箱货船运输，因为这些单元都使用了标准尺寸的集装箱，并且采用了相同的闭锁装置。在泰国进行单元装配和精装修时，这些单元的放置地点已经提前在日本确定了。当货运船到了的时候，这些单元仅仅被卸载下来，放在选定的地点，并且在现场进行快速且经济实用地组装，然后便开始应用了。

场地靠海的一面正对着一个公共公园，为了避免所有的房间都对着这个公园，从而影响公园的气氛，我们将每个单元分开，并且调整每个单元的体积。同时，通过将每个单元分开，使得每个单元都具有良好的隔音效果，在容许的范围内改变墙壁的厚度。建造过程中我们没有焊接连接处，这样我们将来还可以重复使用这些单元。集装箱大小的别墅和附近海边码头停靠的游艇融为一体，显得自然且连贯。

Container-dimension Cottages

Bayside Marina Hotel Yokohama is a project for a seaside cottage hotel. The project uses the standard 40 feet shipping container(approximately 2.4m × 12m) as its modular unit. One unit provides a single "flat" type room, while two stacked units provides the "maisonette" type, with a combined total of thirty-one rooms. By using not only the standard container dimension, but also the same locking hardware, the units were designed to be transported directly by container ship. Assembly and finishing of the units was carried out in Thailand while the site in Japan was prepared in advance. After arrival by ship, the units were simply bolted into place and connected to utilities, allowing for quick and economical on-site completion.

The ocean side of the site faces onto a public park, and in order to avoid having all the hotels windows facing this park and thereby adversely affecting its atmosphere, we separated the units and aligned each volume differently. As well, by separating the units, we could also provide the space necessary to provide a high level of sound insulation at the same time as accommodating a wall thickness restricted by the size the container module. By using non-welded joints in construction we allowed for the possibility of the re-use of the unit in the future. What has resulted is a natural continuity between the arrangement of the units and the ocean view of yachts moored at the adjacent marina.

Twistlock

Anchored for shipping

Diaphragm
St.PL-25 SS400
(Corner casting hole)

St.L-175×175×12

Nut
Washer
Diaphragm
St.PL-25 SS400
(Medium bolt hole)
(Corner casting hole)

Spacer
St.PL-6,SS400
(Medium bolt hole)
(Corner casting hole)

Diaphragm
St.PL-25 SS400
(Medium bolt hole)
(Corner casting hole)

Medium bolt M16

Protective
cap
Double nut

Washer

Diaphragm
St.PL-25 SS400
(Anchor bolt hole)
(Corner casting hole)

Leveling mortar t30

Anchor bolt M12

Continuous foundation

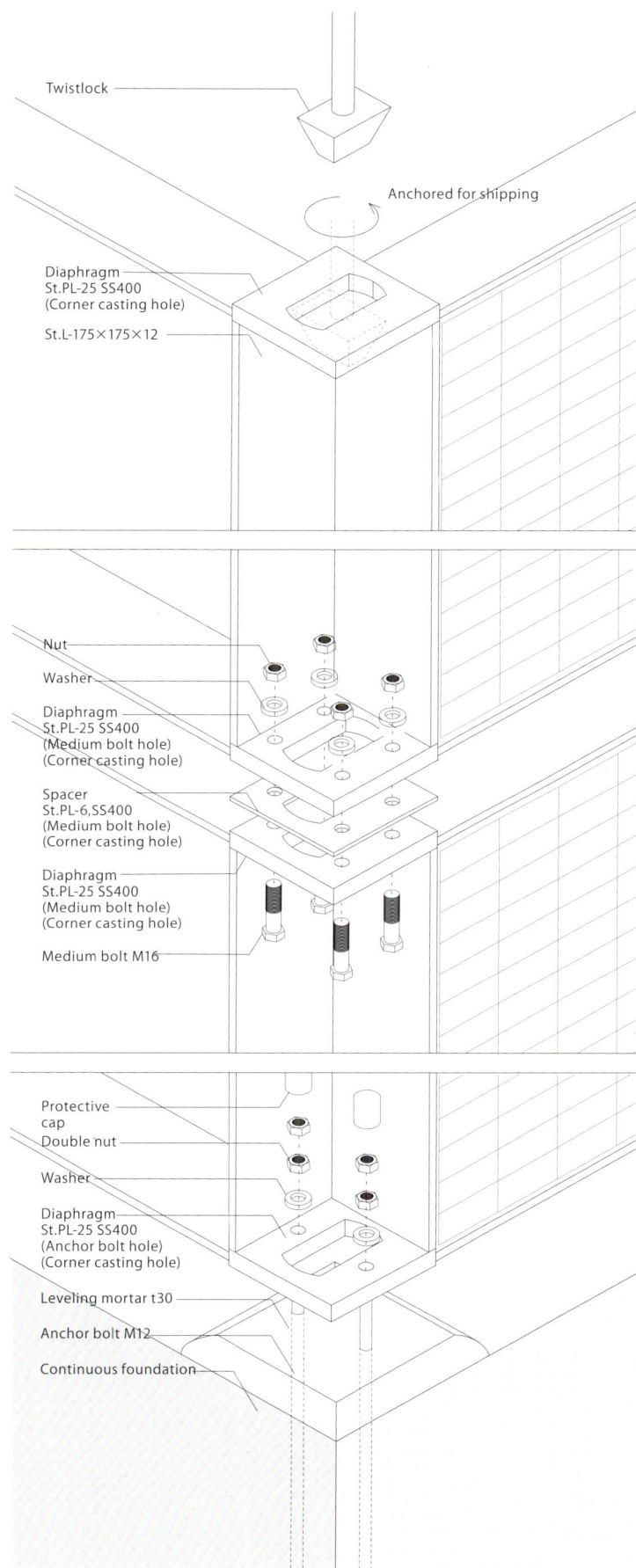

轴测图（转角） Axonometric(corner)

外观5 Exterior V

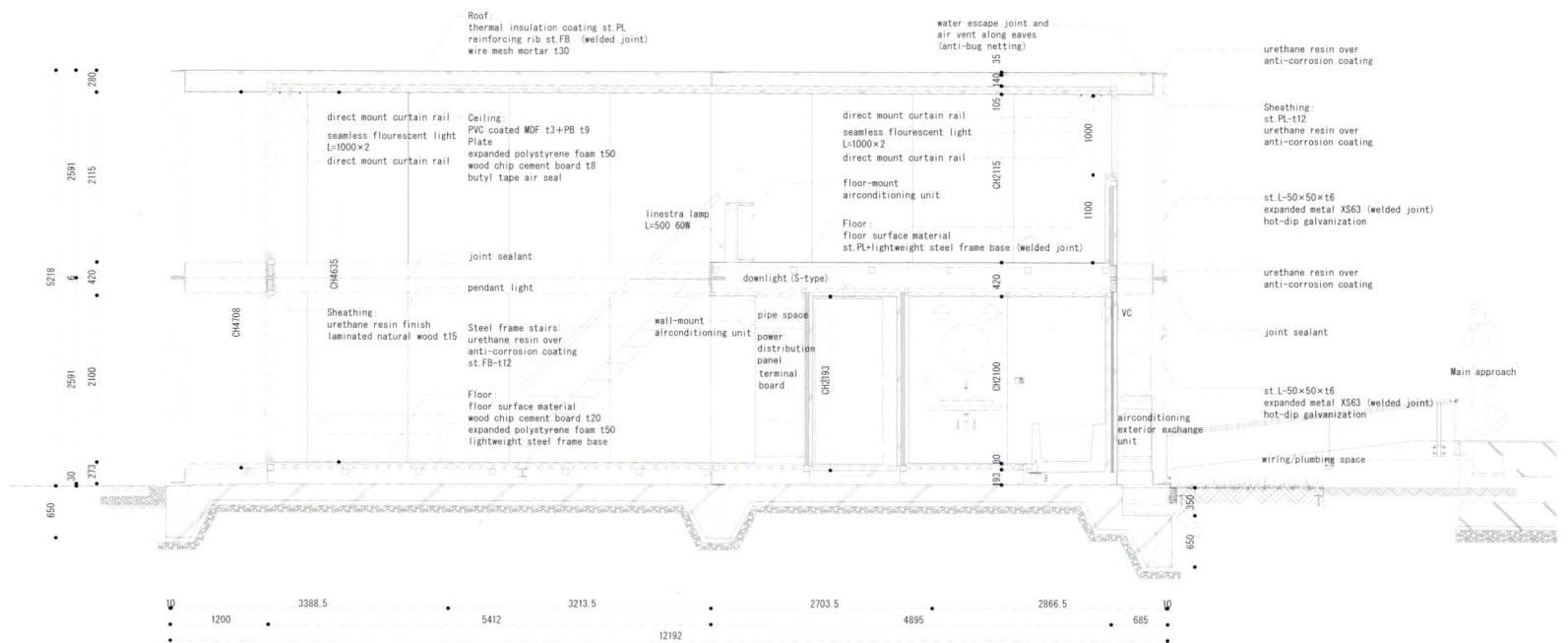

Roof:
thermal insulation coating st.PL
reinforcing rib st.FB (welded joint)
wire mesh mortar t30

water escape joint and
air vent along eaves
(anti-bug netting)

urethane resin over
anti-corrosion coating

direct mount curtain rail
seamless flourescent light
L=1000×2
direct mount curtain rail

Ceiling:
PVC coated MDF t3+PB t9
Plate
expanded polystyrene foam t50
wood chip cement board t8
butyl tape air seal

direct mount curtain rail
seamless flourescent light
L=1000×2
direct mount curtain rail

floor-mount
airconditioning unit

Floor:
floor surface material
st.PL+lightweight steel frame base (welded joint)

Sheathing:
st.PL-t12
urethane resin over
anti-corrosion coating

st.L-50×50×t6
expanded metal XS63 (welded joint)
hot-dip galvanization

linestra lamp
L=500 60W

joint sealant

pendant light

downlight (S-type)

urethane resin over
anti-corrosion coating

Sheathing:
urethane resin finish
laminated natural wood t15

Steel frame stairs:
urethane resin over
anti-corrosion coating
st.FB-t12

wall-mount
airconditioning unit

pipe space

power
distribution
panel
terminal
board

VC

joint sealant

Main approach

Floor:
floor surface material
wood chip cement board t20
expanded polystyrene foam t50
lightweight steel frame base

airconditioning
exterior exchange
unit

st.L-50×50×t6
expanded metal XS63 (welded joint)
hot-dip galvanization

wiring/plumbing space

剖面图（长）Section(long)

Exterior VI

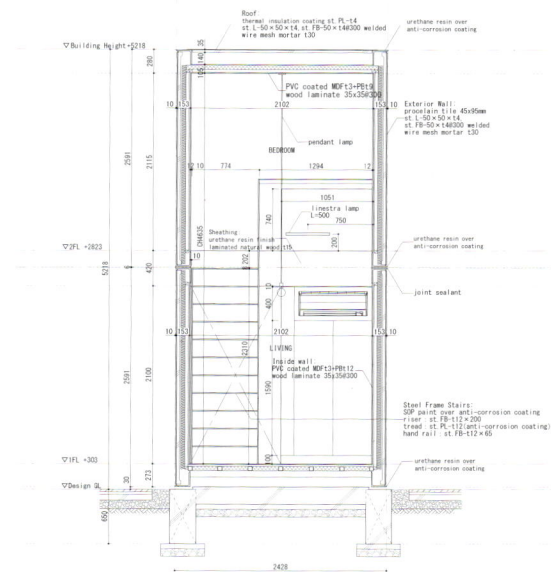

剖面图（短）Section(short)

外观6 Exterior VI

从室内看室外 View from inside to outside

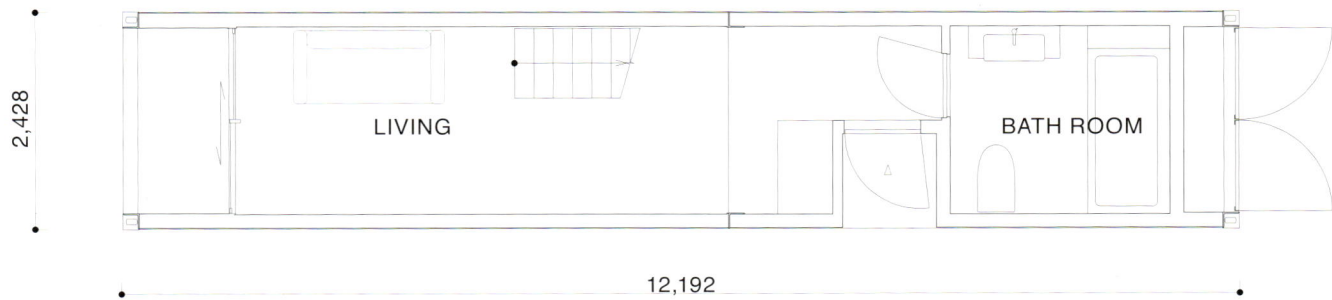

LIVING

BATH ROOM

2,428

12,192

一层平面图（公寓） First floor plan(maisonette type)

室内 Interior

房间 1 Room I

2,428

BED ROOM

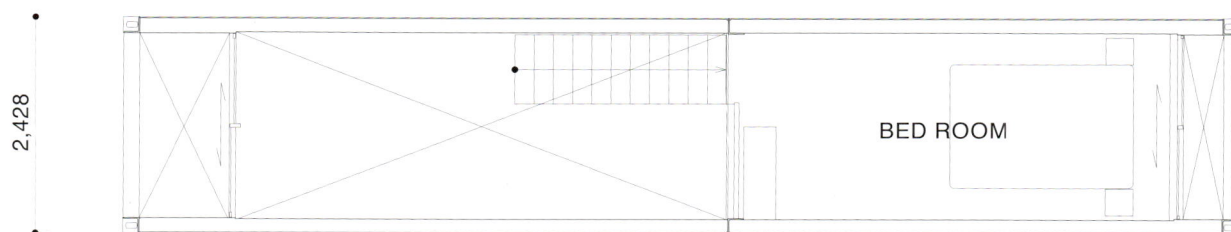

二层平面图（公寓） Second floor plan(maisonette type)

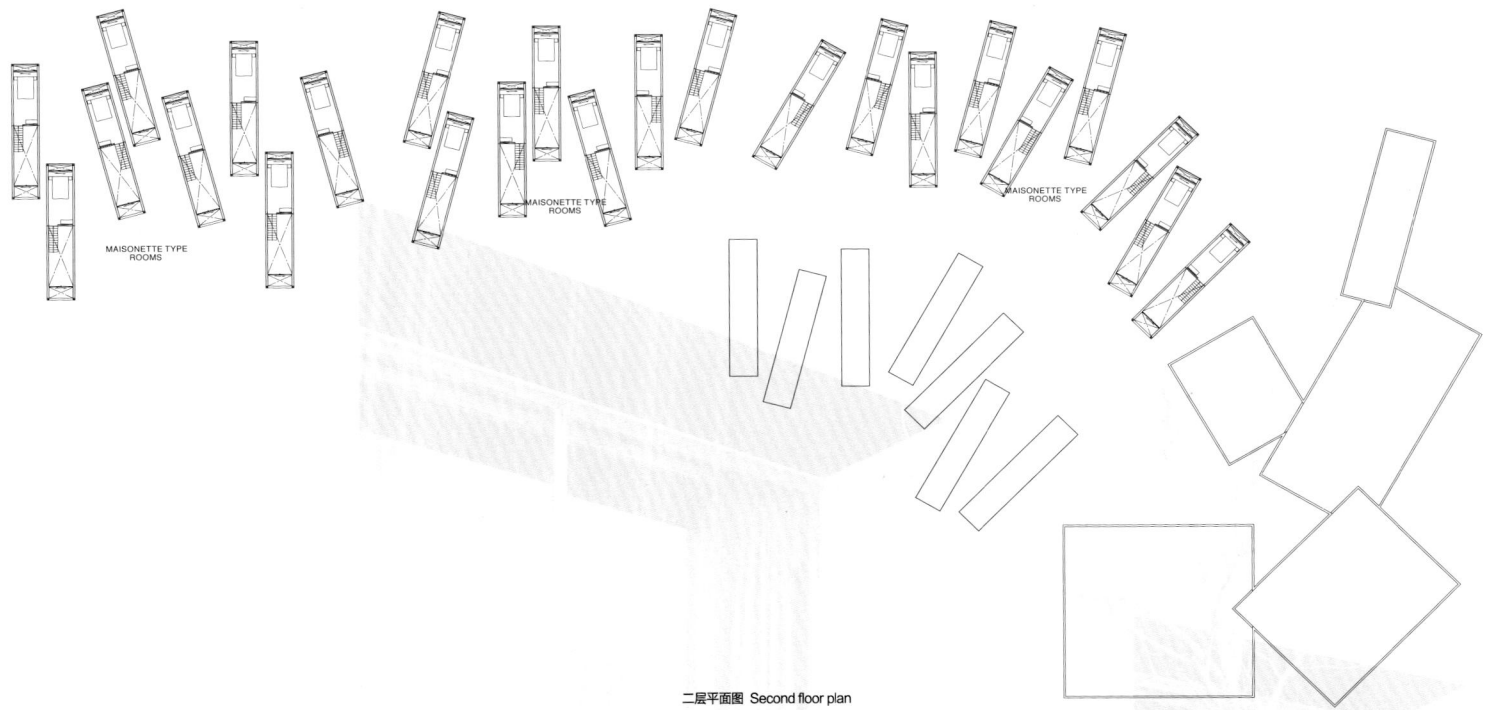

二层平面图 Second floor plan

一层平面图 First floor plan

杜塞尔多夫凯悦酒店
Hyatt Regency Dusseldorf

RIVERSALONS
PEBBLE'S

设计　Sop Architekten
地点　Speditionstrasse 19, Dusseldorf, Germany
室内设计　FG stijl
摄影师　Offered by FG stijl

凯悦酒店是一家全球性的酒店管理公司，并且被公认为是行业的领导品牌和超过 50 年发展历程中改革和创新的代表。杜塞尔多夫凯悦酒店被建设成为 Media Hafen Dusseldorf 的焦点，并且成为登岛的目的地。从远处看，位于公共区域上面高地上的酒吧的鹅卵石般的棚顶，对于游客来说，就是最耀眼最出众的景物。

酒店的入口位于上层房间组成了巨大的悬挑天棚下。紧挨着入口的是一个 26m 宽的具有悠久历史的楼梯。中心的部分将被升起，形成一个顶棚，一个引人入胜的入口可以通向舞厅，在这里可以谈论许多事情。酒店里完全光滑的地面吸引着客人，同时展现着酒店不同的外观。客人可以从主要的入口进入，或者从侧入口直接徒步进入 Dox 酒吧、Dox 饭馆或者 Café D。现代化建筑内部的装饰强调豪华的视觉，并且制造温暖、亲切的感觉。在接待区域，黑色挪威板岩地为公共区域最重要的外观提供了一种良好背景，即金色的箱子。休闲室位于金色箱子的左右方。外面木质的窗和帘使得图书馆内部的气氛很私密并且很温馨。一个火炉使得这样的休息室更像是家。在休息室的对面、金色箱子后面，Dox 酒吧就位于这个地方，它是喝东西的最佳地点。过道和这个酒吧之间用手工钢制的芦苇条装饰，制造出了一种令人兴奋的通视效果。餐馆的第一个区域是露在外面的厨房，这个区域是由露台和中央的白色大理石烤箱组成。在大理石和花岗岩的前面是 5m 长的厨师的桌子，很有特色。在这个桌子的两边，都有两排座位提供足够的私人聚餐空间，其中一排用网状工作窗隐藏，同时具备开放式饭馆的特色。餐馆的第二个区域是水景边的建筑侧面区，就像其他区域一样，这个区域可以在夏天时开放，使饭馆与它的外面平台相联系。第三个餐饮区是 White bar，在这里，厨师在客人面前现场制作寿司。在这个吧中有一面弯曲的后墙，墙上有一些在城市里可以买到的最好的酒，并且可以转换成 DJ 小站。鹅卵石形状的酒吧底部用来作为一个华丽的天花板装饰。一个被巧妙命名为 Pebble's 的酒吧馆成为了这个酒店闪耀的明珠。由抛光后的不锈钢制成的棚顶，闪闪发光，被设计成是这个小岛上重要的特色之一。Pebble's 周围大片的屋顶平台使得客人可以欣赏到水上的整个港口。闪闪发光的外观在室内也可以看到，尤其是地面瓷砖反射的光，当太阳照在上面时，仿佛地上铺上了一层宝石。

水疗房和健身房都坐落在一楼。在健身房里可以看到一个步行桥连接着城市和这个小岛，桥上的灯在晚上闪闪发光，看上去是那么舒服和令人兴奋。水疗区域拥有五个多功能美容和治疗厅，包括一个双人房和一个薇姿治疗室。治疗室最突出的是那些由天然椰子镶嵌成的墙壁，营造出一种暖暖的气氛。在每个治疗室内都有紫水晶装饰。

这个酒店一共有 303 个房间，其中包括 260 间标准间、30 间双人间、10 间初级套房、2 间高管套房和一间总统套房。其最大的挑战之一就是不寻常的构造，这种构造采用了悬梁式设计。巨大倾斜的混凝土柱子安装在建筑物内。创造出一种高雅但是内部实用的设计是令人兴奋的，因为它会创造出令人惊讶的房间，标准间并不十分标准，设计人员试图把屋子的前半部分设计成一个奢华试衣间和放松的区域。在这个区域后面的第二个区域，床对着窗户，从而欣赏到壮丽的景色。这种景色同样可以从浴室内看到，由于浴室和卧室是玻璃隔开的。在这些套房里，视野是设计的重点：无论客人坐在客厅时，还是在床上休息时，或者在化妆台前时，都可以看到水边和城市的壮丽景色。总统套房内有奢华的钢琴、两个壁炉和小镇中最好的视野，因为它位于建筑的最宽处，朝着步行桥。

Hyatt is a global hospitality company with widely recognized, industry leading brands and a tradition of innovation developed over their more than fifty-year history. The Hyatt Regency Dusseldorf, set to act as focal point and destination on the tip of the island in the Media Hafen Dusseldorf. From a distance the bar pavilion Pebble's, situated on the plateau above the public spaces, will be the shining recognition point for all visitors.

The entrance to the hotel is under the enormous cantilevered canopy created by the rooms above. Next to the entrance is a 26 meter wide monumental staircase. A central section can be lifted to form a canopy: a spectacular entrance to the ballroom which is set to be the talk of many events. The completely glazed ground floor of the hotel entice guests showing the different aspects of the hotel. Guests can enter either through the main entrance or through the side entrances along the promenade directly into Dox bar, Dox restaurant or Café D. The interior is based on emphasizing this magnificent view and on creating warmth and intimacy in a modern building. The black Norwegian slate floor at the reception area creates the perfect background for the most dominant aspect of the public spaces, the Golden Box. Behind the Golden Box on the left the Lounge is situated. The wooden screens on the outside emphasize the library like feeling inside, creating privacy and a warm atmosphere. A fire place makes this lounge feel like home even more. Opposite of the lounge, behind the Golden Box, Dox bar is located, the perfect place for drinks. Handmade steel reeds are used as decoration between the aisle and the bar, creating an exciting see-through. The first area of the restaurant is the show kitchen with its back drop of the patio and central white marble ovens. In front of the marble and granite an almost 5 meter long chef's table acts as a central feature. On either side of this table, two booths, half concealed with lattice work screens, provide ample spacefor more intimate dinner parties within the context of an open restaurant. The second area of the restaurant is the waterside facade which, like several others, can easily be opened to let the restaurant flow out to connect with its exterior terrace during summer. The third dining area is the White bar with sushi chefs preparing food in front of the guests. This bar has a curled back wall showing some of the finest wines available in the city and can also be transformed into a DJ-booth. The bottom of Pebble's bar serves as a shining ceiling decoration. The jewel of the hotel is the bar pavilion aptly named Pebble's. This shimmering river bolder shaped pavilion of polished stainless steel is set to shine as a central feature on the tip of the island. The large roof terraces around Pebble's enable guests to enjoy the view of the entire harbor form above to the water. The shimmer and shine of the outside is also to be found inside the bar, especially in the mosaic tiles on the floor, which is turned into a mere jewel when the sun shines on it.

The spa and gym are located on the first floor. The gym has an excellent and inspiring view on the pedestrian bridge connecting the island to the city, which is lit at night. The spa area contains five multifunctional beauty and treatment rooms including one double room and a Vichy treatment room. The focal point in each treatment room is the mosaic wall of natural coconut, creating a warm surrounding. In each treatment room a purple amethyst-crystal is placed centrally.

The hotel contains 303 guestrooms, including 260 king rooms, 30 twin rooms, 10 junior suites, 2 executive suites and 1 presidential suite. One of the biggest challenges was the unusual structure of the building because of the cantilevered design. Throughout the building enormous diagonal concrete pillars have been installed. It was exciting to create an elegant but practical interior design, resulting in surprising rooms. The standard king room is not very standard. In this new concept the first half of the room is designed as a luxurious dressing and pampering area. Behind this area in the second half of the room, the bed is facing the window to enjoy the spectacular view. This view can be seen from the bath tub as well, because of the glass between the bath tub and the bedroom. In the suites the view was the focal point for the design: whether a guest is seated in the living area, in bed, in the bath or in front of the make-up table, the view over the water and the city is spectacular. The presidential suite is complete with grand piano, double fireplace and the best views in town positioned as it is on the entire width of the buildingfacing the pedestrian bridge.

外观 1 Exterior I

外观 2 Exterior II

White 酒吧 White bar

Dox 酒吧 Dox bar

休息区 Lounge

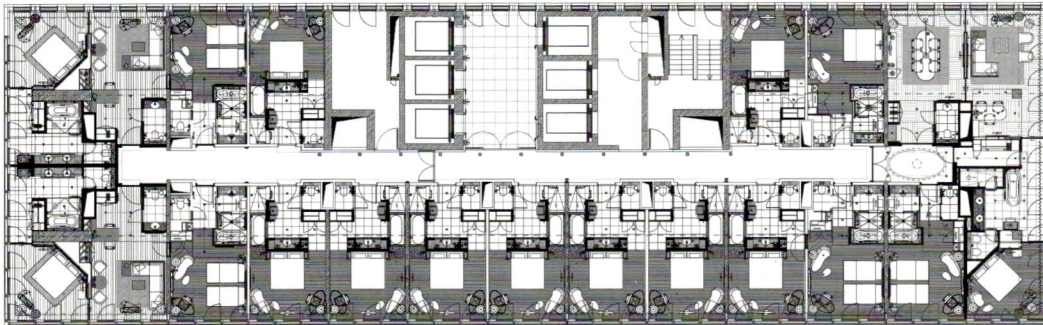

十五层平面图 Fifteenth floor plan

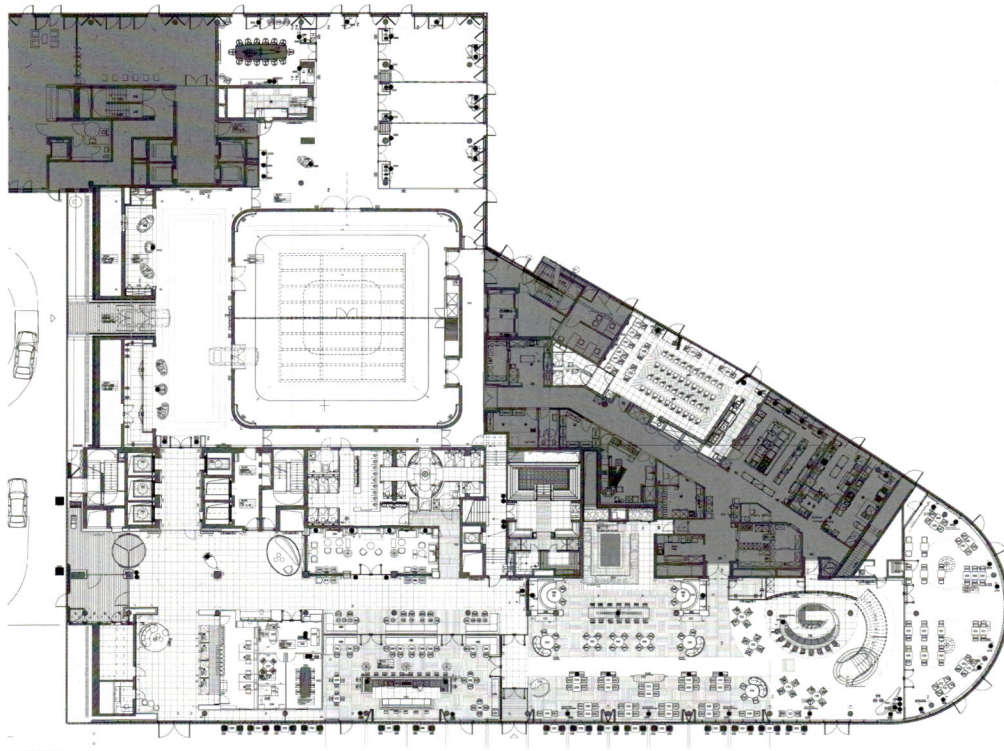

一层平面图 First floor plan

休闲室 Lounge

套房 Suite room

凯悦酒店俱乐部 Regency club

未来酒店
Future Hotel

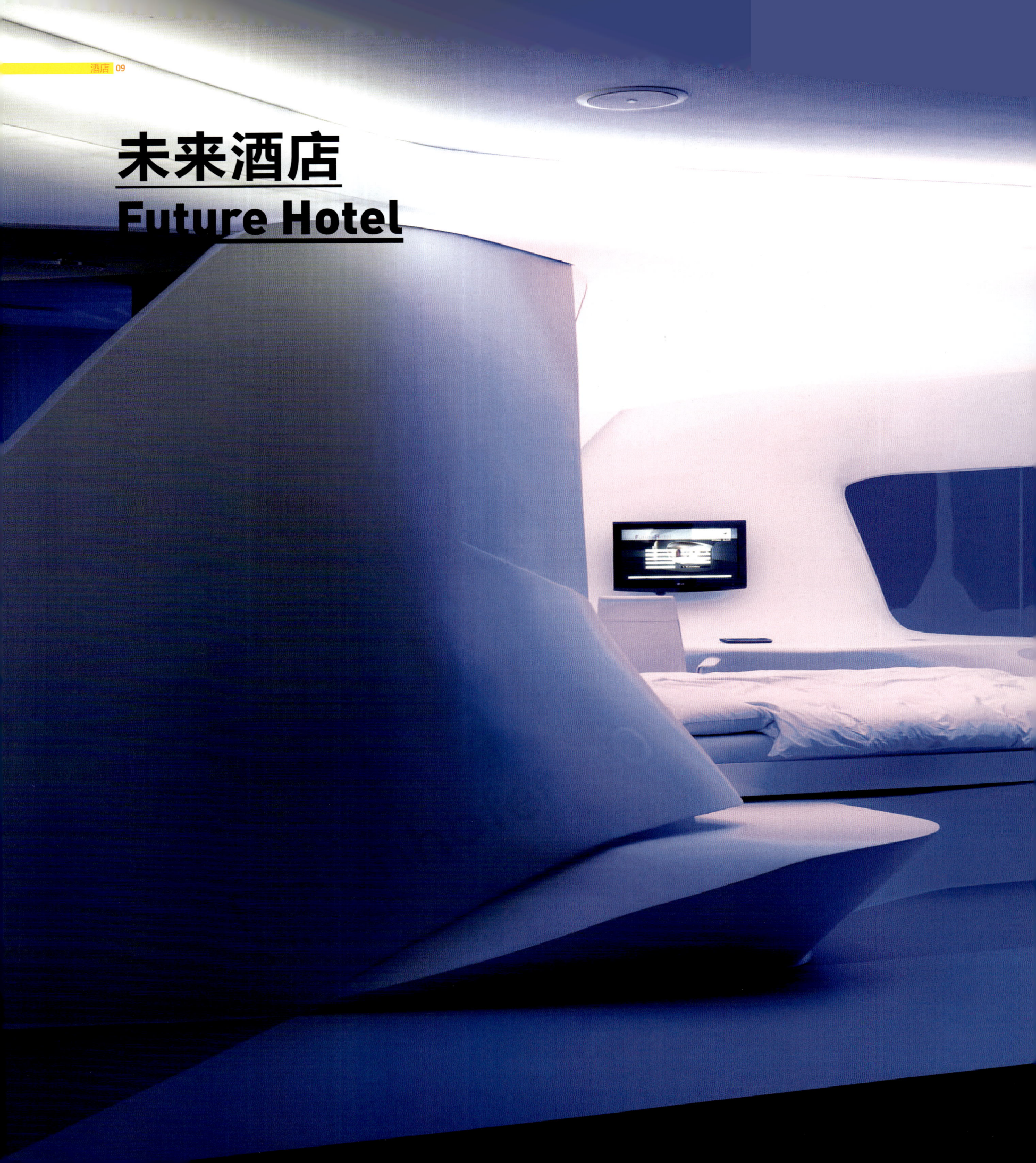

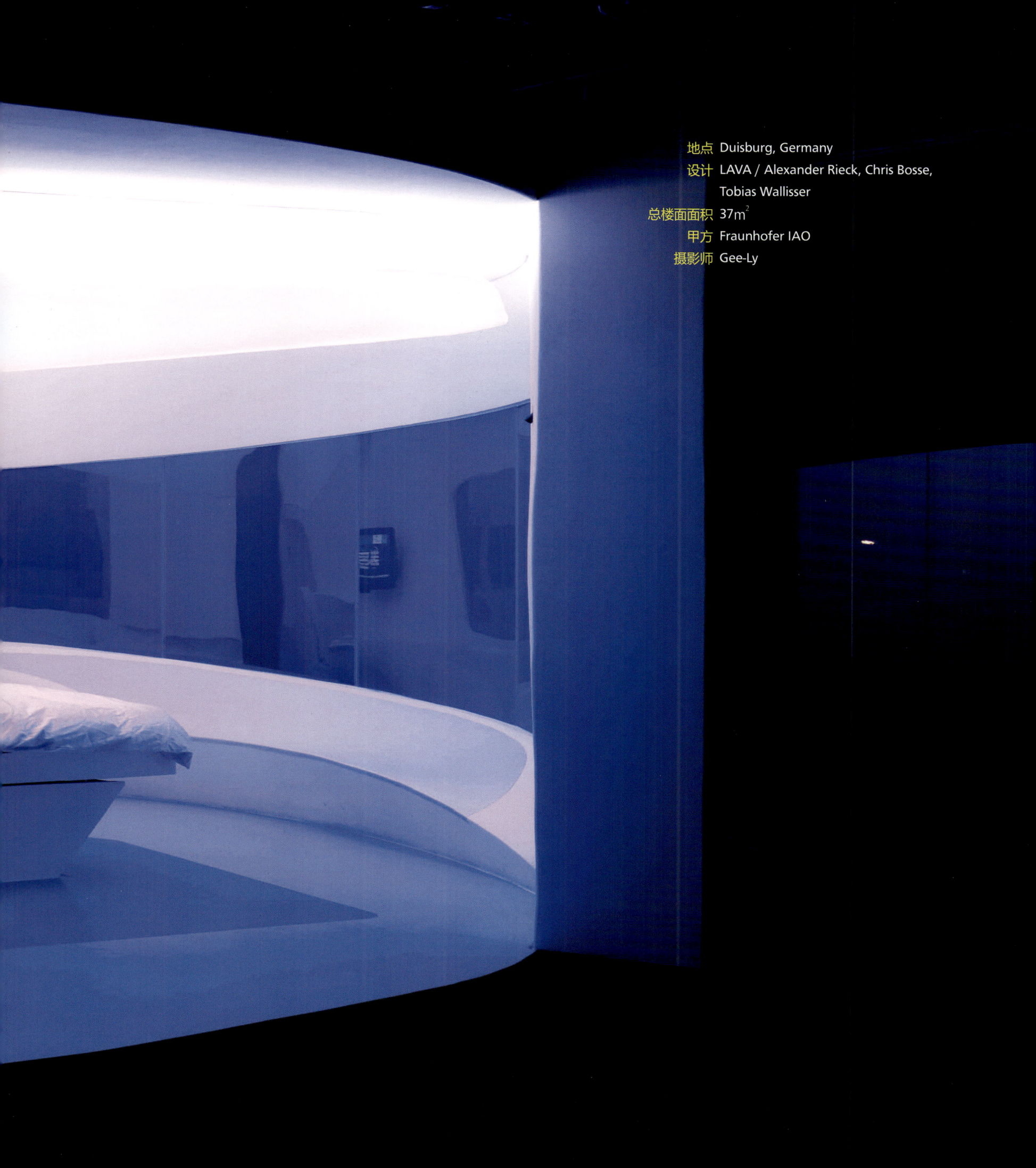

地点 Duisburg, Germany
设计 LAVA / Alexander Rieck, Chris Bosse, Tobias Wallisser
总楼面面积 37m^2
甲方 Fraunhofer IAO
摄影师 Gee-Ly

magicmirror

bath
4 sqm

1.2 sqm

wc

shower

3.4 sqm

4.3 sqm

wardrobe
1.9 sqm

2 sqm

futurespa

OLED light

bedtable

bed

chair

table

19.9 sqm

couch

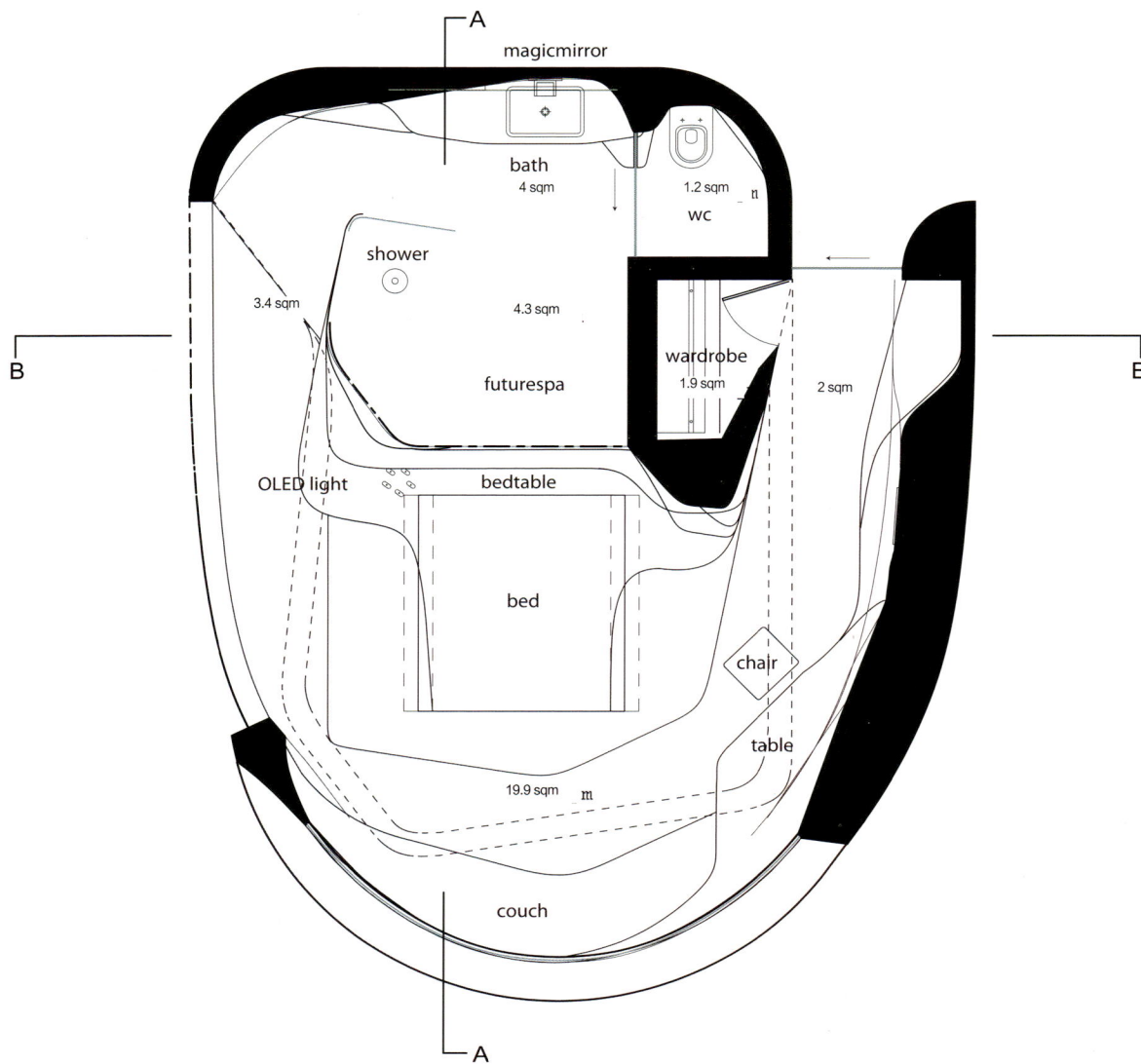

楼层平面图 Floor plan

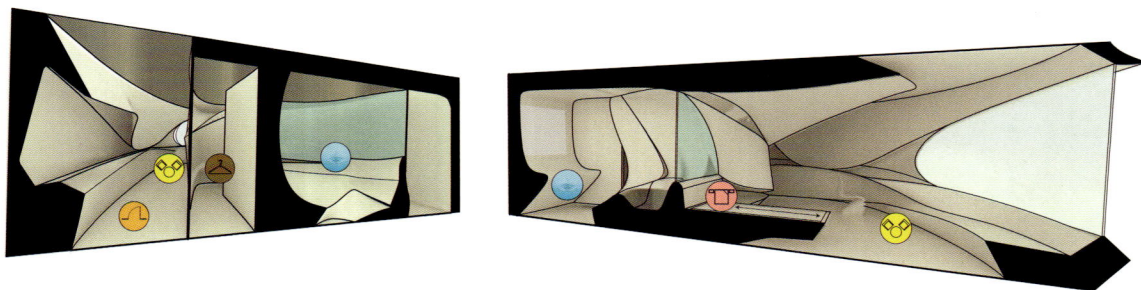

轴测图 Axonometric

卧室全貌 *Overview on the bedroom*

Construction photos

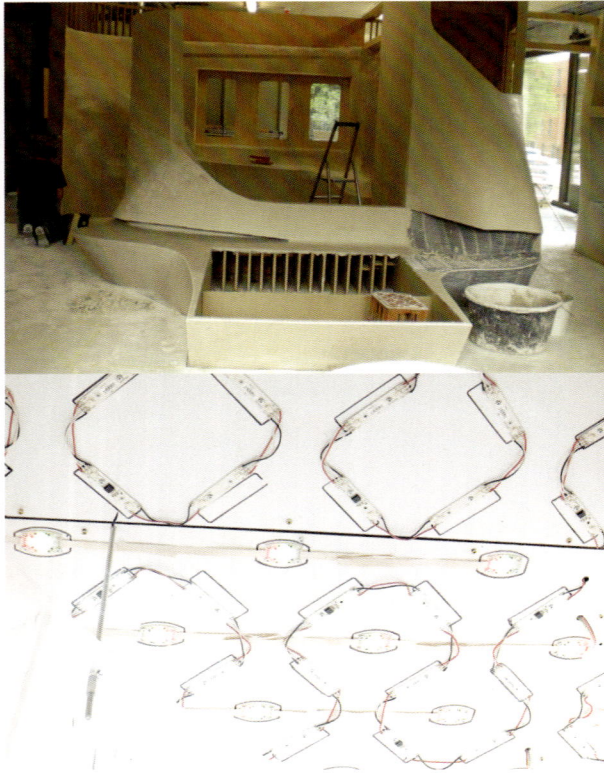

当代室内装修实例

我们设计了未来酒店样板间，这是一项研究建筑和技术创新之间接口的示范项目。通过采用参数化设计方法和半自动作业，及时地实现了初始设计。整个建筑过程没有使用任何传统工程图。数字化模型和实际建造的样板间之间不存在任何差距，完全实现了设计规划。项目合作方之间的紧密合作为我们带来了很多新见解，这将会应用在我们在阿联酋将来的酒店项目中。因此，未来酒店样板间的远景能比预期更早成为现实。

项目哲学和方法

未来酒店关注的是使用未来科技来满足酒店住客的期待和要求。样板间淡化了技术和内部空间之间的区别，在媒体和通讯方面采用最新的技术，并结合使用知名厂商开发的原型产品。人体的舒适感是我们考虑的最主要因素，而在背景中起作用的技术则几乎看不出来，却为人们提供了单独控制媒体、光线和气候的机会。创新部分包括反时差灯、动态而舒适的床、个人水疗区、智能化的镜子和大型媒体展示窗。我们设计了一个空间连续体，将所有区域集中到一个单一的形态中。自由形式的和谐式外表面承载着基础设施，成为技术和人体间的分界面，其主要特点是由奇数的边缘来突出的流动性过渡。软材料和硬材料的结合产生了功能方面高度协调的区别。

材料和细部设计目的

整个样板间使用了三种材料——石膏板、木材和透明的膜——这与安静的环境协调一致。白色的石膏板被调整为下部结构的形状，该结构按照数字化模型的设计，采用CNC技术打磨。使用石膏板能够创造出流动的、光滑的和锐角的表面。个人水疗区主要由木材构成，以便创造出一个更温暖舒适的环境。此处也采用锐角边和喷射软边。整个天花板几乎都用透明的膜覆盖着，这种膜同时起到照明的作用。通过改变LED灯的颜色及其在周围白色的墙上留下的不同效果，可以调节不同的气氛。不同材料之间的转变通过嵌缝边上精细的细部设计完成——嵌缝边界定了房间形状和功能。

Contemporary interior design practice

We have designed the Future Hotel Showcase Room, a demonstration project that researches the interface between Architecture and technological innovation. Applying parametric design methods and semi-automated production allowed for an intime realisation of the original design. The whole building process was done without any conventional technical drawings. The gap between digital model and the built showcase was inexistent and led to an identical realisation of the design. The collaboration between the project partners generated many new insights, which will be implemented in our upcoming hotel projects in the U.A.E. The vision of the future hotel showcase could thus become reality sooner than expected.

Project philosophy and methodology

The Future Hotel focuses on meeting the expectations and requirements of hotel guests using tomorrow's technology. Blurring the definition between technology and interior space, the showcase room features the latest technology in the fields of media and communication in combination with prototypical products developed by renowned manufacturers. Human comfort was of paramount concern, while technology functions almost invisibly in the background, providing the opportunity for individual control of media, light and climate. Some of the innovations are anti-jet-lag lights, an active comfort bed, a personal spa area, an intelligent mirror and a large media display window. We designed a spatial continuum, integrating all areas into one single gesture. The free-formed harmonic outer skin takes up the infrastructure and becomes the interface between technology and the human body, characterised by fluid transitions accentuated by singular edges. A combination of soft and hard material creates a well-balanced differentiation of functional aspects.

Material and detailing intent

The whole showcase consists of three main materials – plasterboard, timber and translucent membrane – which harmonize into a calm environment. The white plasterboard is adapted to the shape of the under construction, which was milled using CNC techniques according to the digital model. The use of plasterboard made it possible to achieve fluid, smooth and sharp edged surfaces. The personal spa area is mainly made out of timber, in order to achieve a more warm and comfortable surrounding. Also here is played with sharp but jet soft edges. Almost the whole ceiling is covered with translucent membrane, which functions as a lighting element. Different moods can be achieved through changing the color of the LED lights and thereby also the white surrounding walls. Transition from one material to the other is made by fine detailing over a spline – edges which define the shape and function of the room.

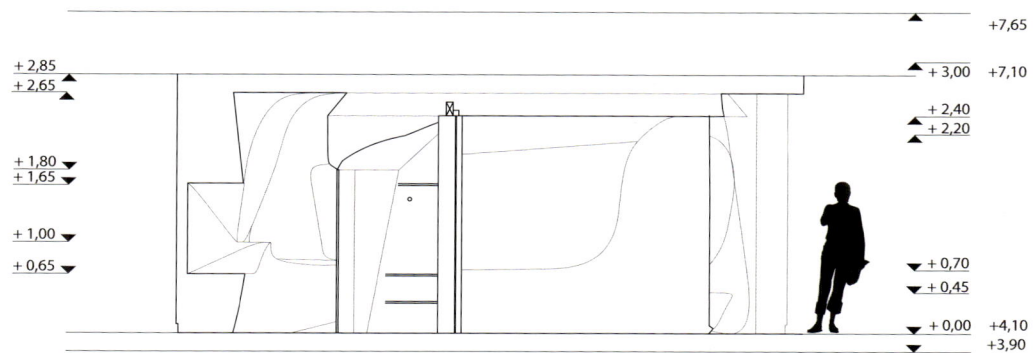

剖面图 B－B Section B－B

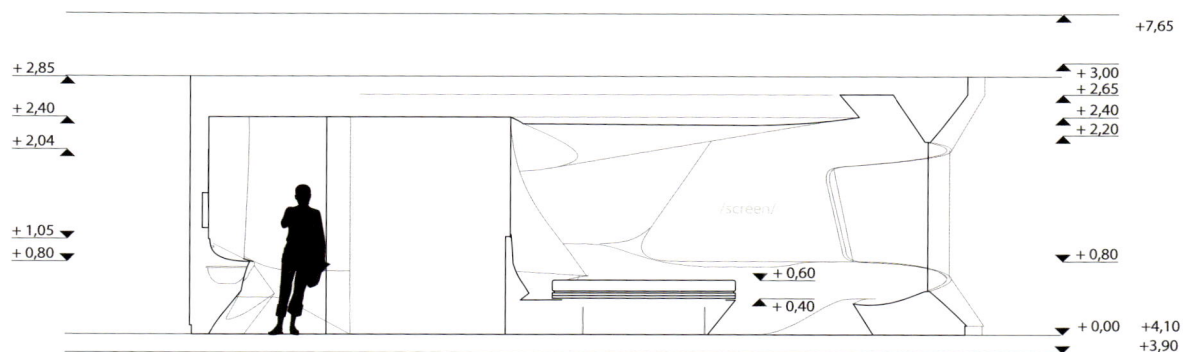

剖面图 A－A Section A－A

浴室 Bathroom

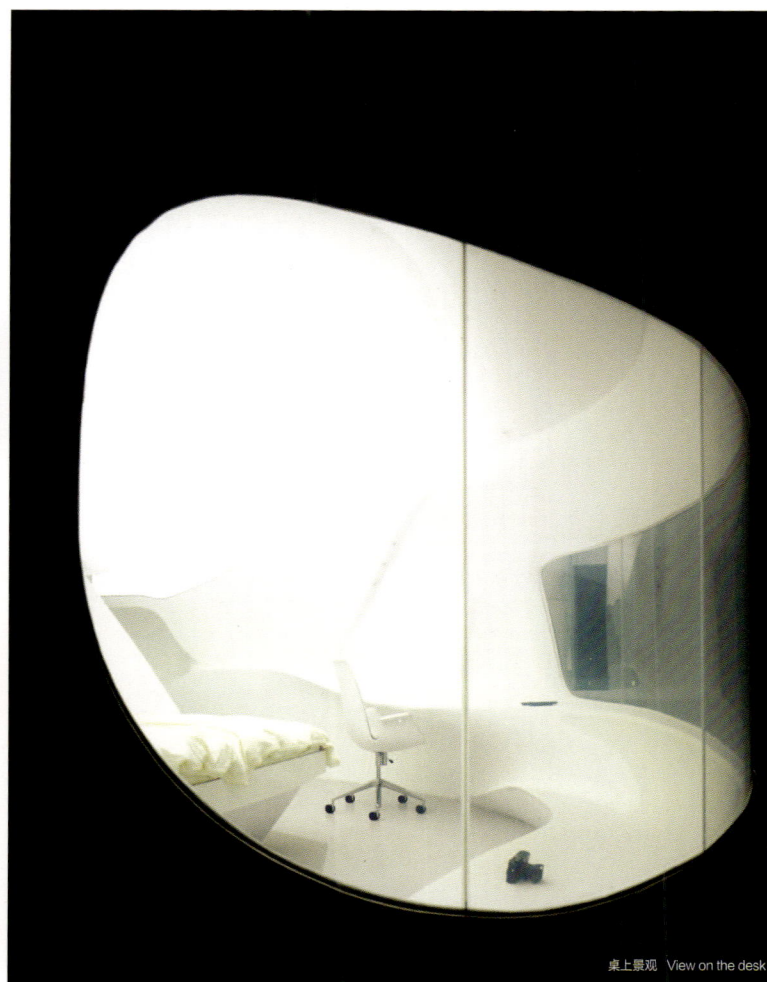

桌上景观 View on the desk

墙及入口细部　Detail on the wall & the entrance

墙及窗户细部　Detail on the wall & the window

卧室全貌 Overview on the bedroom

海岬酒店
Headlands Hotel

地点 Austinmer, Australia
建筑面积 10,000m²
参与设计人员 Jarrod Lamshed, Anh-Dao Trinh, Jonas Epper, Niklas Muehlich,
Andrea Dorici, Alessio Coghene
甲方 George Dimitrovski

0   50   100   150   200   250

总平面图 Site plan

海岬酒店开发项目是一个国际标准的酒店资产计划，已经吸引了众多国际知名的酒店集团的主要酒店品牌的兴趣。海岬酒店的项目投资和运营前景很乐观。海岬酒店的地区性需求经过调研，并在政府委任的经济研究中做了概述，这些经济研究旨在预测酒店与旅游业的当地和地区性指导原则。现有的当地客户群表明支持海岬酒店开发，使其将大量的当地特色和历史带入海岬的下一阶段。海岬酒店对 Austinmer 来说代表着一个巨大的机会，使其成为沿着新南威尔士海岸线的沿海旅游业的一个地标性建筑，是一个著名的呼应社区、环境可持续和商业可行的开发项目。海岬酒店独特的沿海位置和它与悉尼大都会区域的公路和铁路网络极其相似的特点，使旅游者很容易接触到该地区原始的自然美景。当地的边界为：北面和西面是一个多山的皇家国家公园，东面是南太平洋，周围还有新南威尔士第三大城市伍伦贡的经济活动区。计划的开发项目包括 120 个酒店房间、40 个豪华居民公寓、水疗间、公共酒吧、餐馆、咖啡馆、舞厅和集团会议设施，所有这些都包含在一个 10 000m² 的楼层面积之内。海岬酒店提案建立在通过分析和了解该位置全面开发潜力而被获准的一个开发提案之上。当前计划的开发规模经过计算和结算，使其满足公认酒店集团的品牌和运营要求。项目所在地独特的机遇，加上建筑设计的状态和最近集合起来的专业开发团队，为投资者和经营者等提供了一个投资这一特殊的澳大利亚沿海酒店资产的极好机会。

The Headlands Hotel development proposed an International standard hotel asset that has already attracted major Hotel Brand interest from internationally respected hotel groups. The projected investment and operational outlook for the Headlands Hotel is positive. Regional demand for the Headlands Hotel product has been researched and outlined in government commissioned economic studies that forecast the local and regional direction of the hospitality and tourism sector. An existing local clientele have indicated support for the development headlands hotel that will carry the extensive local character and history into the headland's next phase. The Headland represents a spectacular opportunity for Austinmer to become a landmark of coastal hospitality known for community responsive, environmentally sustainable and commercially viable development along the New South Wales coastal edge. The unique coastal location of the Headlands, with its close proximity to Sydney Metropolitan area road and rail network, allows visitors easy access the regions pristine natural beauty. The local area is bordered by the a mountainous state owned Royal National Park to the north and west, the South Pacific ocean to the east and the economic activity of Wollongong, the 3rd largest City in New South Wales. The planned development consists of 120 Hotel Rooms, 40 residential Luxury Apartments, Spa Rooms, Public Bar, Restaurant, Café, Ballroom and Corporate Conference facilities, all contained within an allocated floor area of 10,000 m². The Headlands Hotel proposal builds upon a currently approved Development Proposal by analyzing and realizing the sites full development potential. The currently proposed development size has been calculated and balanced to fulfill brand and operational requirements of recognized hotel groups. This site unique opportunity, coupled with state of the architectural design and a currently assembled specialist development team, offers investors and operators alike an exceptional opportunity to capitalize on this distinctive Australian Coastal Hotel asset.

G   L1   L2   L3   L4   L5   L6

option 1

option 2

option 3

legend

Car park
BoH hotel
Lobby hotel

Bar
Restaurant
Function Room

Day spa
Hotel rooms
Luxurious serviced apartments

+6   LUXURY RESIDENTIAL ACCOMMODATION

+5   LUXURY RESIDENTIAL ACCOMMODATION

+4   HOTEL ROOMS

+3   HOTEL ROOMS

+2   HOTEL ROOMS

     KITCHEN

+1   HOTEL ROOMS

     SPA, FITNESS AND HEALTH FACILITIES
     CONVENTION BOARD ROOM
     DOUBLE HIGH BALLROOM
     MULTIUSE RESTAURANT SPACE

0   MULTIUSE RESTAURANT SPACE
     PUBLIC BAR
     HOTEL ADMINISTRATION
     LOBBY - LOUNGE
     60 UNDER GROUND CAR PARK

-1   120 UNDERGROUND CAR PARK

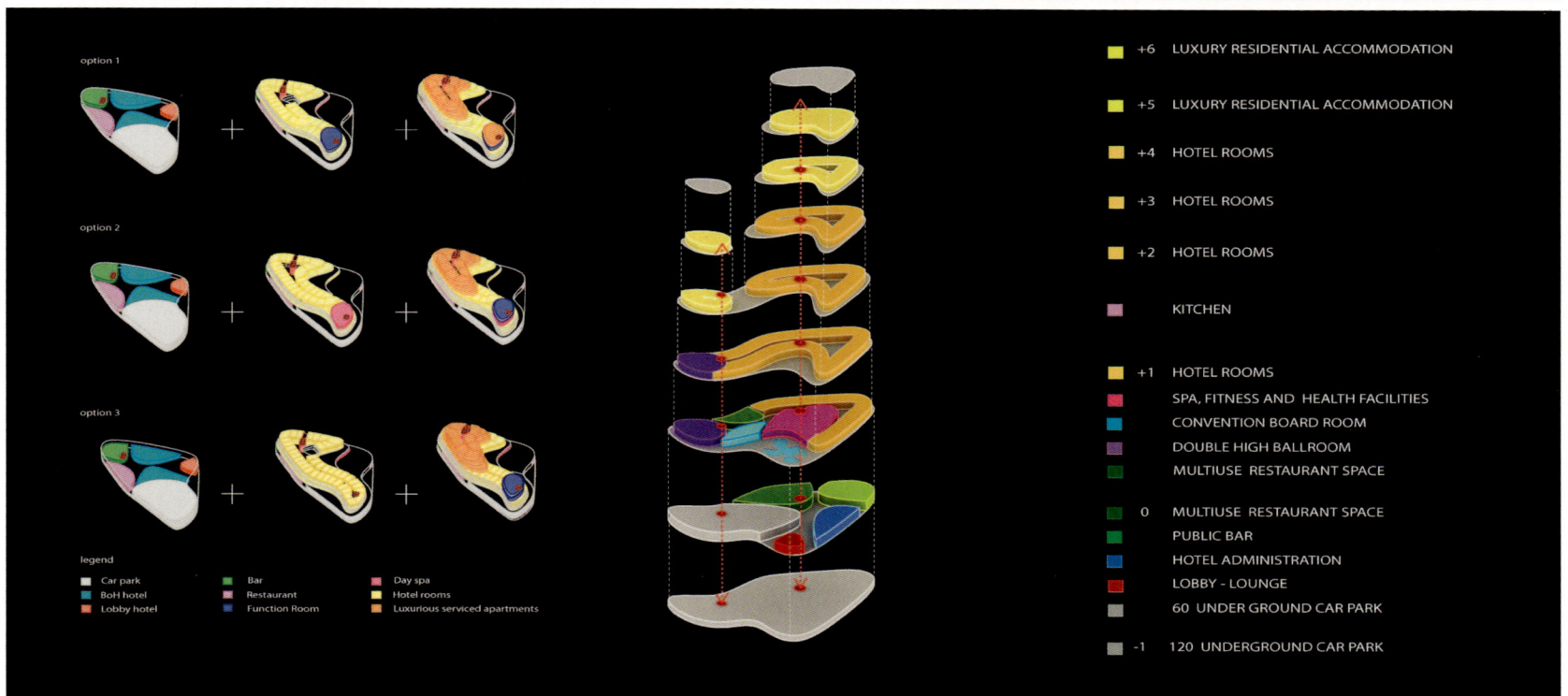

示意图 Diagram

模型 1 Model I

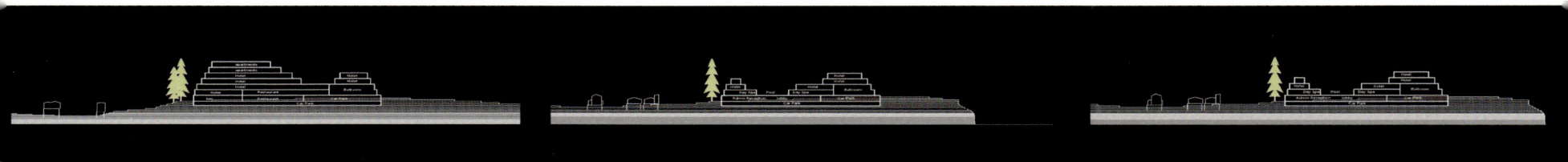

剖面图 Section

室外 Exterior

模型 2 Model II

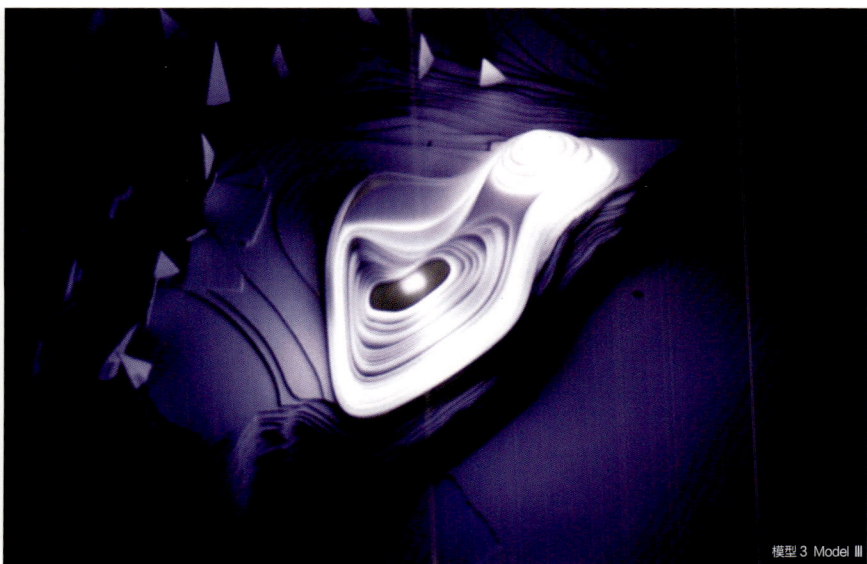

模型 3 Model III

观景楼酒店
Belvedere Hotel

地点　Mykonos, Greece
设计　Rockwell Group / David Rockwell
基地面积　9,000m^2
参与设计人员　owner_Nikolas, Tasos, Domna Ioannidis, principal_Edmond Bakos, Shawn
　　　　Sullivan, staff_Ioannidou, Jean Marc Tang, Katie Putnam, Thom Ortiz, Sally
　　　　Weinand, Issei Summa(rockwell group), Aktor SA(consultants & engineering)
　　　　Johnson Light Studio / Clark Johnson(lighting)
甲方　Belvedere Hotel
摄影师　declared in picture

© Ed Reeve

外观 1 Exterior 1
©Ed Reeve

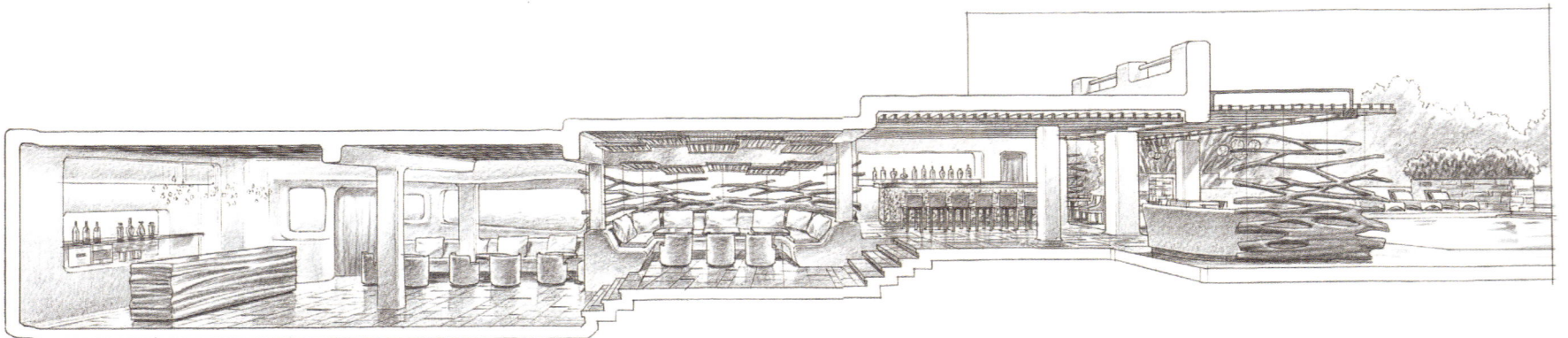

剖面图 Section

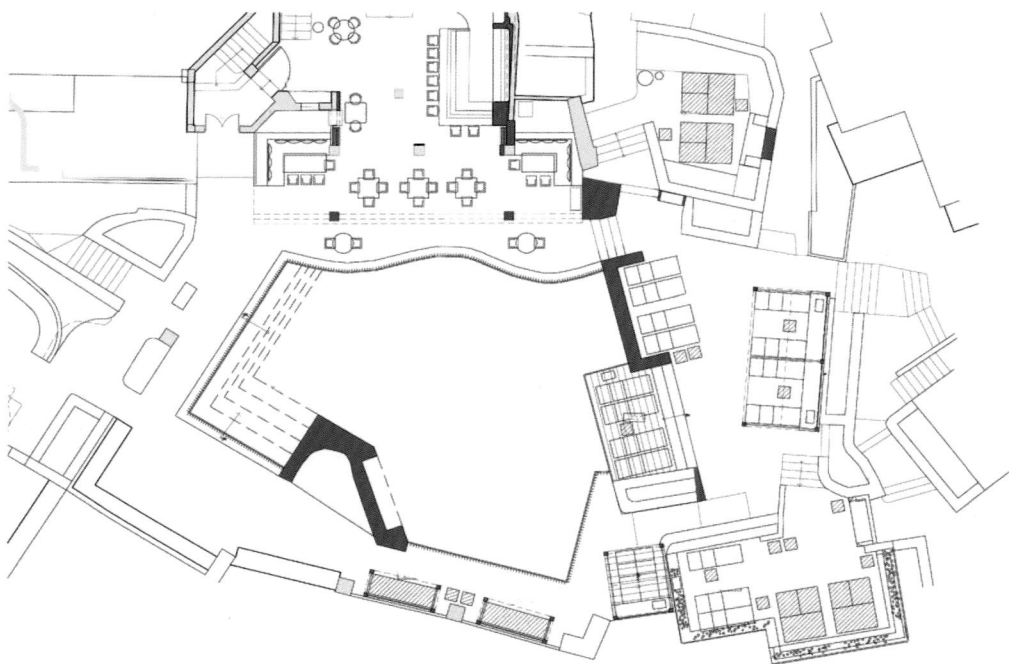

楼层平面图（室外） Floor plan (exterior)

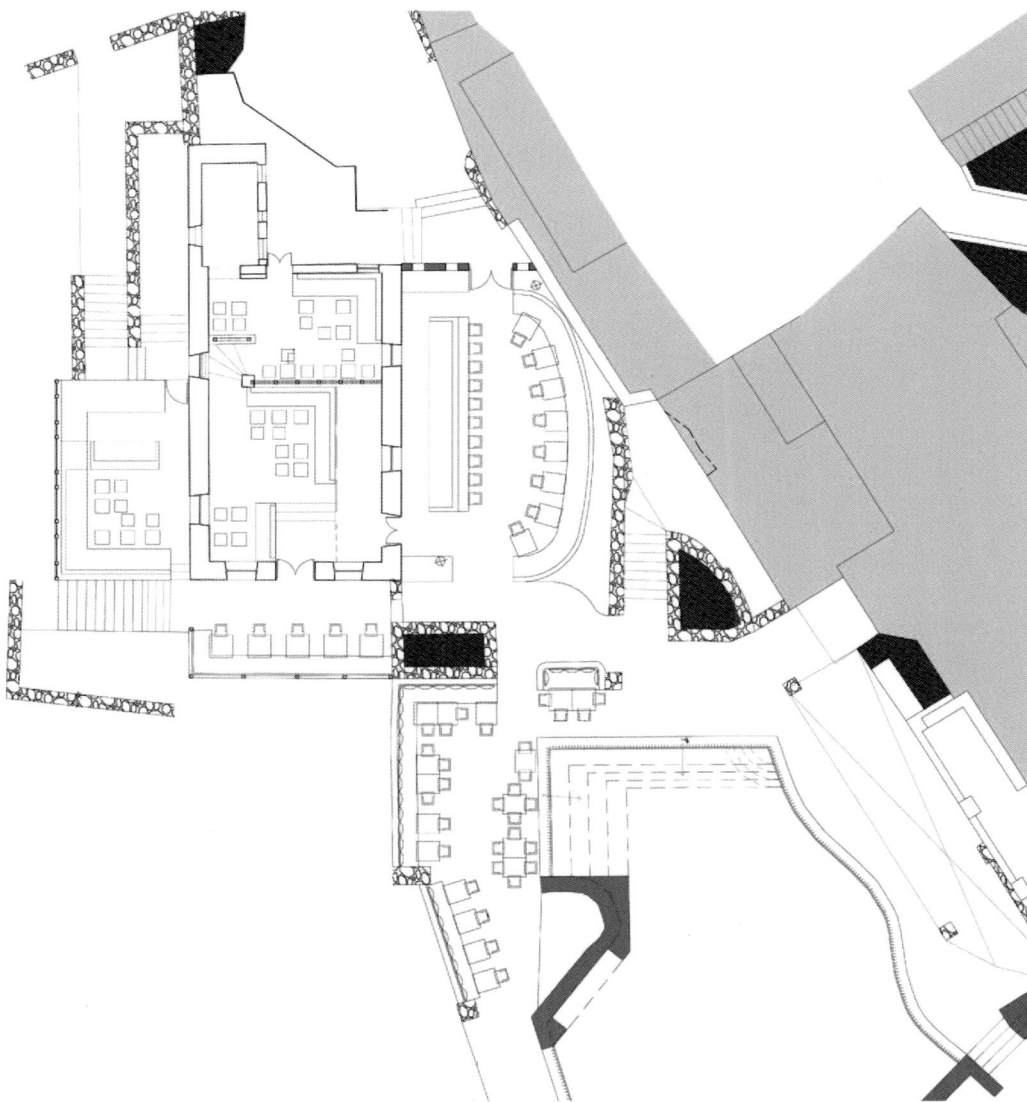

楼层平面图（matsuhisa） Floor plan (matsuhisa)

室外 / Exterior (night)
© Vagelis Paterakis

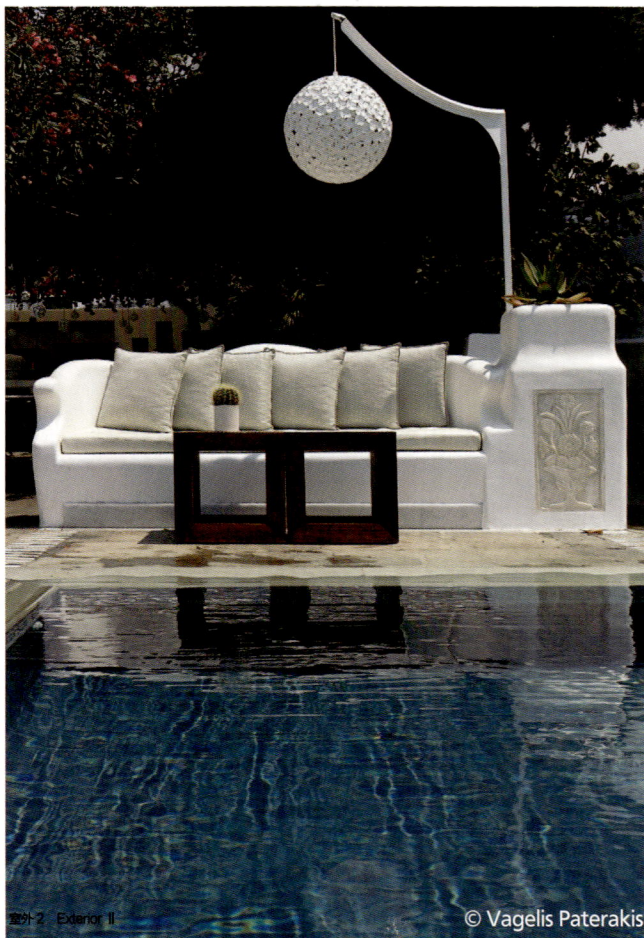
室外2 Exterior II
© Vagelis Paterakis

观景楼酒店餐厅和酒吧

观景楼酒店餐厅和酒吧的设计受到爱琴海的启发，是一个三层的空间结构。使用的材料和设计元素也效仿了源于海洋的元素。

客人从顶层的酒吧进入，酒吧由灵活边缘的木质吧台台面构成，这个台面由大理石吧台前脸支撑。不锈钢压铆螺母柱上的手雕大理石覆盖着吧台的门脸，并从后面打光，形成悬浮着的效果。

上层休息室位于酒吧下层，在两侧位置是弯曲的抹灰人行道，界定了房间的形状。手雕的风化木屏像极了带有空隙和曲线的水波纹，这些木屏位于人行道的后面。下层主要就餐区的中心也有这种木屏。

下层是餐厅和一个附属酒吧。酒吧有一个木质的吧台表面，上面雕刻的图案代表水的流动。主餐厅的墙用白色石膏覆盖。雕刻的壁龛镶嵌着珍珠母。地面由带有金属镶嵌物的石料制成。由印花金属模制品构成的马格达莱纳球形灯闪闪发光，灯光的效果类似于在水下看到的太阳的效果。

Belvedere Hotel Restaurant and Bar

Inspired by the Aegean Sea, the Belvedere Hotel Restaurant and Bar is a three-level space. The materials and design elements used echo those that are native to the sea.

Guests enter on the top level, where is a bar is comprised of a live edge wood bar top supported by a marble bar face. Hand-carved marble stones on stainless steel standoffs cover the bar face and are backlit to create the appearance that they are floating.

Located on the level below, the upper lounge is flanked on two sides by curved plaster banquettes that define the shape of the room. Hand-carved weathered wood screens that resemble rippling water with voids and curves are located behind the banquettes. These screens also appear on the lower level in the center of the main dining area.

The restaurant and an additional bar are located on the lower level. The bar has a wood bar face that is carved to represent the movement of water. The walls of the main dining room are covered in white plaster. Carved-out wall niches are lined with mother-of-pearl inlay. The floor is made from stone with metal insets. Magdalena Globe lights, which are made of molded pieces of stamped metal, emit a glittering light that is similar in effect to the sun seen underwater.

© Ed Reeve
游泳池景观　View from swimming pool

© Ed Reeve
外观 3　Exterior III

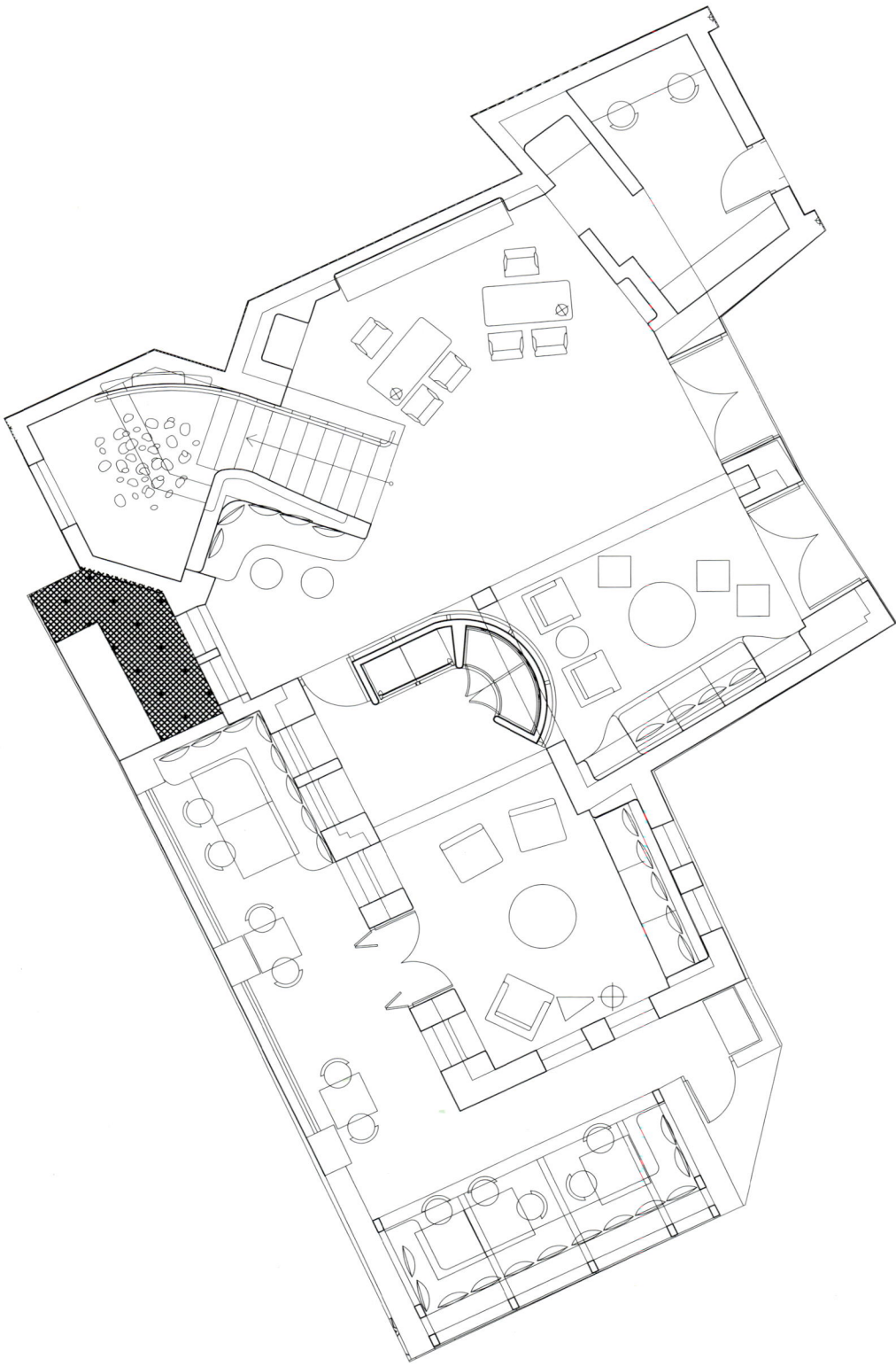

楼层平面图（大厅） Floor plan (Lobby)

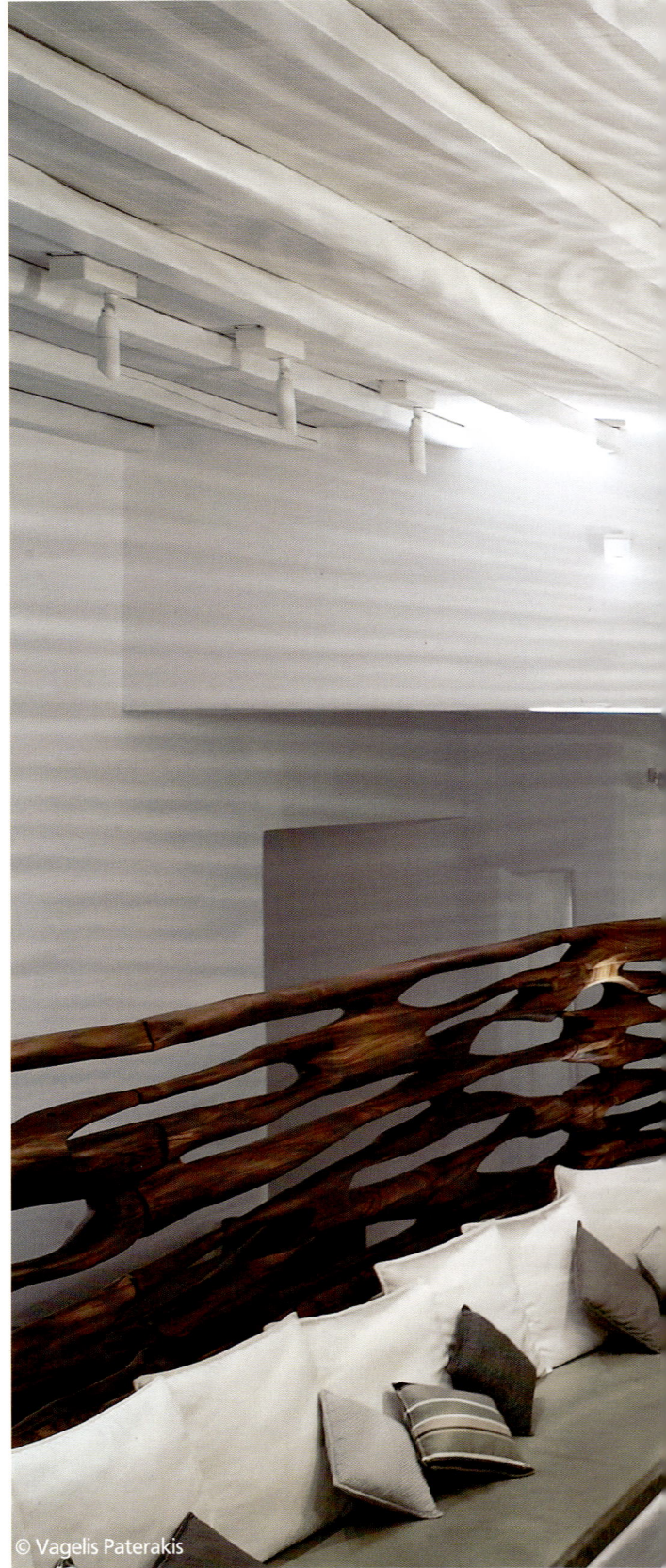

© Vagelis Paterakis

大厅 Lobby

套房 Suite

© Ed Reeve

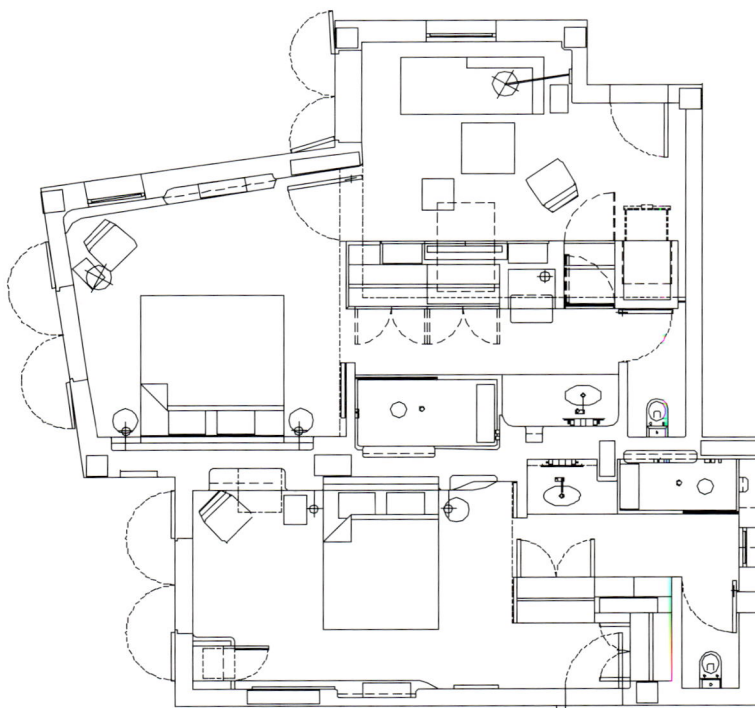

楼层平面图（小套间和标准房间） Floor plan (junior suite & standard room)

© Ed Reeve

主卧室浴室 Master bathroom

楼层平面图（主套房） Floor plan (master suite)

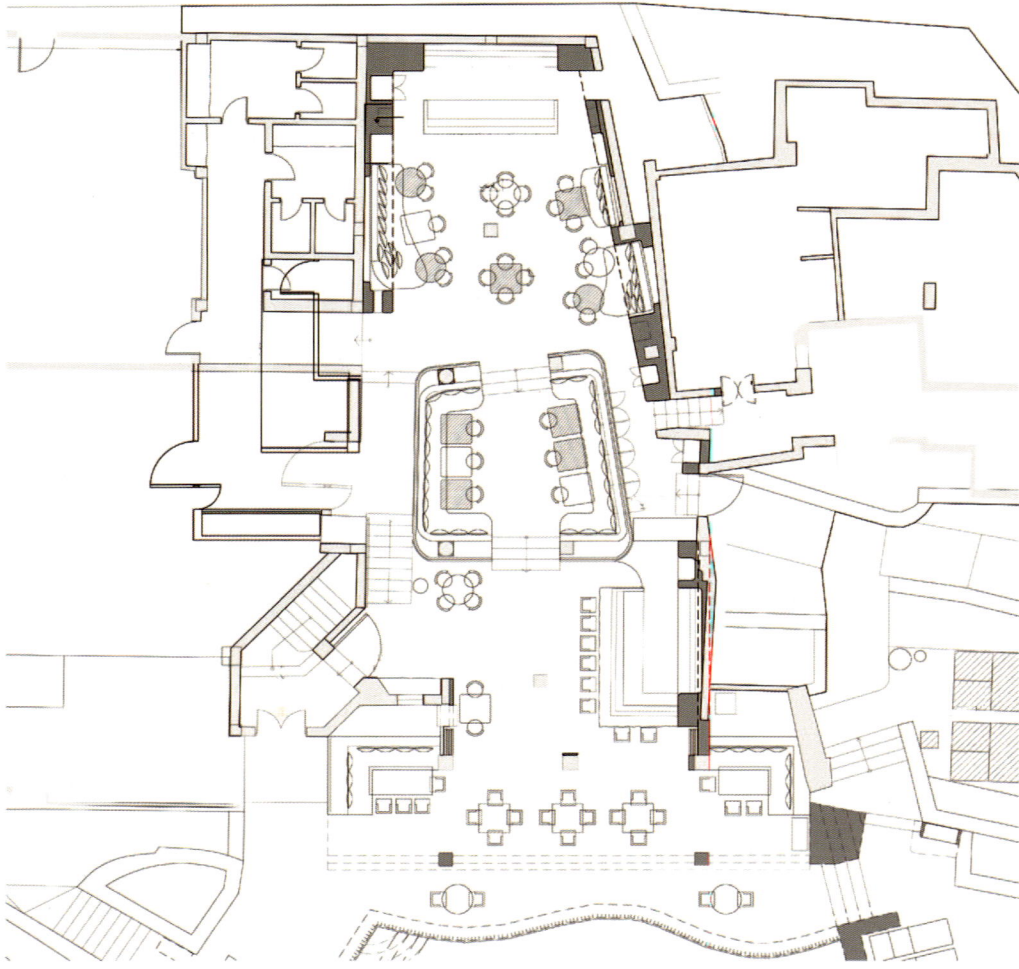

楼层平面图（室内） Floor plan(interior)

© Ed Reeve

© Ed Reeve

艾姆斯酒店
Ames Hotel

地点 Boston, America
设计 Rockwell Group / David Rockwell
总楼面面积 70,000m^2
参与设计人员 Gregory Stanford, Jessica Davenport, Charles Farruggio(Rockwell Group),
Mari Balastrazzi, Heather Maloney, Tracy Smith(Morgans Hotel Group),
ADD INC., C7 Architects, consultant_Tillet Lighting
甲方 Normandy Partners and Morgans Hotel Group
摄影师 offered by Morgans Hotel Group

大厅 1 Lobby 1

设计理念

尽管全新的内部设计保留了对原有 19 世纪外部设计的尊重，罗克韦尔团队在大厅、伍德沃德餐厅和 113 个翻新的客房的设计里，使用了现代的框架结构。原有的元素也被重新诠释，完善了优雅并具有创新性的新设计。

设计细节

大厅的特色是由原始的大理石马赛克瓷砖铺成的拱形天花板、引人注目的大理石和黄铜楼梯。大厅周围散布着原有场地专属的艺术品，例如一盏高悬在棚顶的枝形吊灯，灯上有数千个反光的光盘，用线悬挂在地板上方。接待处的后面有一个抽象的陶瓷艺术幕墙，是由许多片手工制造的瓷器组成的。

大厅后面的两层酒店餐厅也是有着时代特征对比鲜明的地方，将厚重的历史背景和现代气息极好地结合在一起。这个餐厅叫作伍德沃德，以年历的作者纳撒尼尔 艾姆斯曾拥有过的小酒馆命名。每一层都有一个主要的吧台和一个正式的餐饮区，每处饰面都有着精巧而又出乎意料的特点，这一点受到裁缝为人量身定做衣服的启发。餐厅的主要特点是二层的"珍品储藏室"，一个维多利亚风格的架子贯穿整个餐饮厅，架子上摆放着超过 200 件精心挑选的艺术品。高光白色饰面的温莎椅零散地摆放在酒店里，使由木镶板构成的墙面如同沐浴在乳白的瓷漆里。

Design Concept

Although the design for the all-new interior of the space pays homage to the original 19th century exterior, Rockwell Group provides a modern framework for the lobby, Woodward restaurant, and 113 renovated guest rooms. Original elements are re-interpreted to complement the elegant and innovative new design.

Design Details

The lobby features an original marble mosaic tile vaulted ceiling, and a dramatic marble and brass staircase. Scattered around the lobby are original site-specific art works such as a chandelier of thousands of reflective discs suspended over the floor on wires, and an abstract ceramic wall installation behind the reception area made up of many pieces of hand cast porcelain.

Off of the lobby the two-story hotel restaurant is also a contrast of eras, merging the historical context with a touch of the modern day dandy. It is called Woodward, named for a tavern once owned by Almanac author Nathaniel Ames. Each floor features a main bar as well as a formal dining area, with finishes inspired by bespoke men's tailoring with subtle unexpected details. The main feature of the restaurant is the "Cabinet of Curiosities" on the second floor, a Victorian-inspired shelf throughout the dining space curated with over 200 hand selected objects of inspiration. Windsor chairs, epoxied in high gloss white finish, are scattered throughout the restaurant, complimenting the wood paneled walls washed in a milky enamel.

大厅 2 Lobby II

客房 1 Guest room |

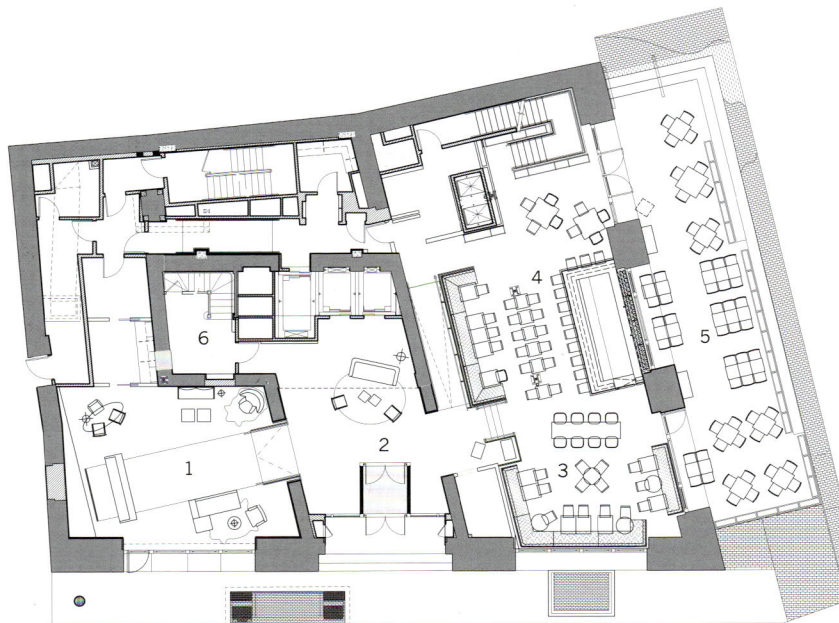

一层平面图 First floor plan

客房 2 Guest room II

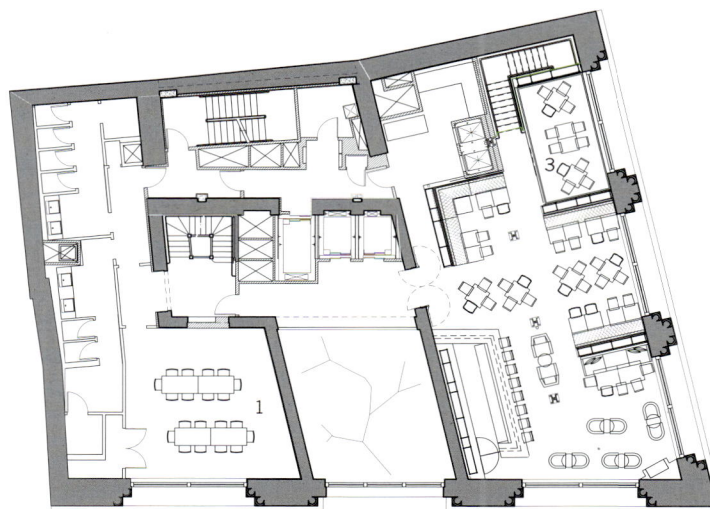

二层平面图 Second floor plan

室内 1 Interior I

室内 2 Interior II

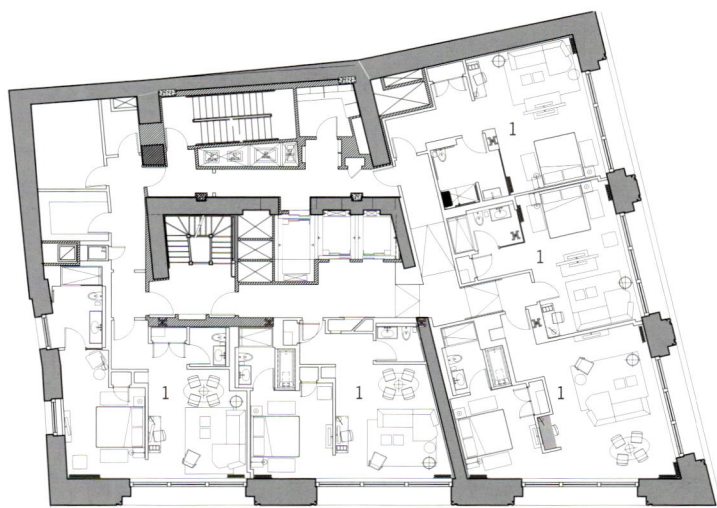

三层平面图 Third floor plan

The visitor who reaches Boston, indeed by water, can hardly fail to be struck with the natural beauties - heightened now by artificial adornment - of the harbor, narrowing, as it does, in even curves on either side, dotted with many turf and undulating or craggy islands; - long stretches of beach being visible almost to the horizon, now and then interspersed by a jutting, cliff-bound promontory, or pushing out seaward a straggling, shapeless peninsula of green.

Picturesque America, 1874

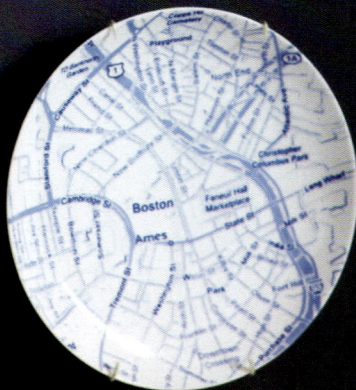

Utoco 深海治疗中心与酒店
Utoco Deep Sea Therapy Center & Hotel

设计	Ciel Rouge Creation / Henri Gueydan, Fumiko Kaneko
地点	Muroto Misaki, Japan
功能	Therapy center, Hotel
基地面积	35,545m^2
总楼面面积	3,071m^2
结构	Reinforced concrete
饰面材料	Exposed concrete, Plaster board, Aluminium sash, Aluminium, AEP painting, Coating, Tile, Carpet(int.)
摄影师	Toshihisa Ishii

入口 Entrance

从走廊向室外看 View from corridor to outside

Utoco 深海治疗中心与酒店点缀在青山与白浪构成的和谐的大自然中。这座酒店以对健康的狂热追求而闻名，这一点通过舒适的、寻求自然的设计来寻求舒适的理念体现得淋漓尽致。形状和形式胜过其精神：曲线令人感到舒适，圆圈使人安心，水平的线条使所有感官都平静下来。这些由简单的、普通的、令人放松的、纯粹的原始图形，如椭圆形和圆形构成的形状，与贝壳、鱼腹或海湾形状的空腔巧妙地融合在一起。这座酒店和海水浴治疗中心接纳了一种新的文化，这种文化与深海海水惊人的力量有关，回应了人们对健康日益觉醒的、不断提高的关注。只有简单的设计，没有额外的装饰，这个空间给顾客提供了一个放松的环境。每间客房都用白色和红色来装饰，昏暗的灯光强调了房间的空间感，营造了一个宁静的气氛。Utoco 深海治疗中心与酒店用独特的形状和简单的图像吸引人们，使自己成为一个完美的休息空间。

Utoco Deep Sea Therapy Center & Hotel is located in nature where green mountains and white turbulent waves are harmonized. The hotel, which focuses on health craze, presents comfortableness through the natural design. Shapes and forms transcend to the spirit: curves are comforting, circles reassure and horizontality calms down the senses. The shapes are made of simple, pervasive, relaxing, of pure origin as ovoid and circles, gently melting with soft cavities in shell-like, fish belly or bay forms. This Hotel and Thalasso Therapy Center comes to embrace a new culture linked to the phenomenal powers of deep see water and answering the demand for the growing awareness in wellness concerns. With simple design without excess, the space provides the customers with relaxing environment. Each guestroom is clad in white and primary red color, and dim illumination emphasizes the space, creating a calm atmosphere. Utoco Deep Sea Therapy Center & Hotel attracts people with the extraordinary shape and simplified image, establishing itself as a perfect space for rest.

外观 Exterior

鸟瞰图 Bird's eye view

游泳池 1 Pool I

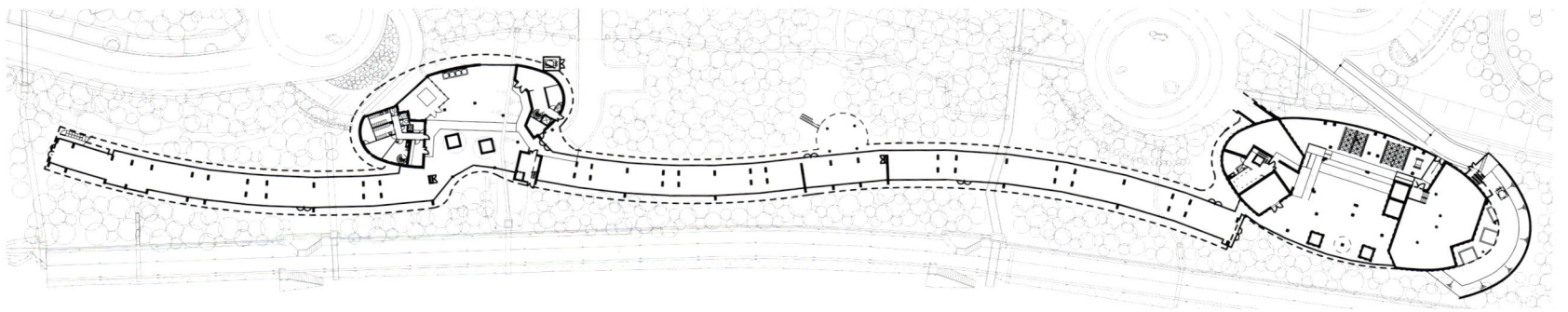

一层平面图 First floor plan

游泳池 2 Pool Ⅱ

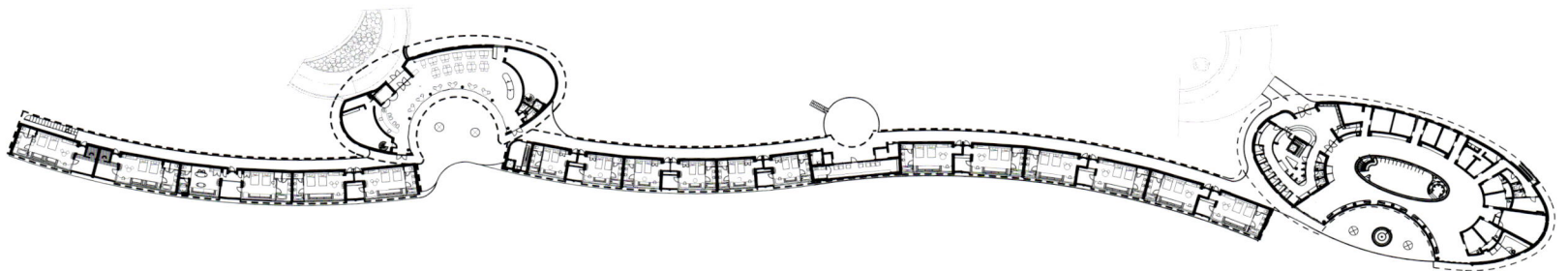

二层平面图 Second floor plan

厨房（夜景）Kitchen (night)

厨房 Kitchen

室内 Interior

浴室 Bathroom

房间 Room

仁川喜来登酒店
Sheraton Incheon Hotel

设计　Hok. Inc., USA
　　　Di Leonardo International, USA
　　　Heerim Architects & Planners Haeahn Architecture. Inc.
地点　Incheon-si, Korea
基地面积　10,234m²
建筑面积　3,198m²
总楼面面积　60,959m²
景观面积　1,617m²
楼层　B3, 22FL

西北视图 Northwest view

入口 Entrance

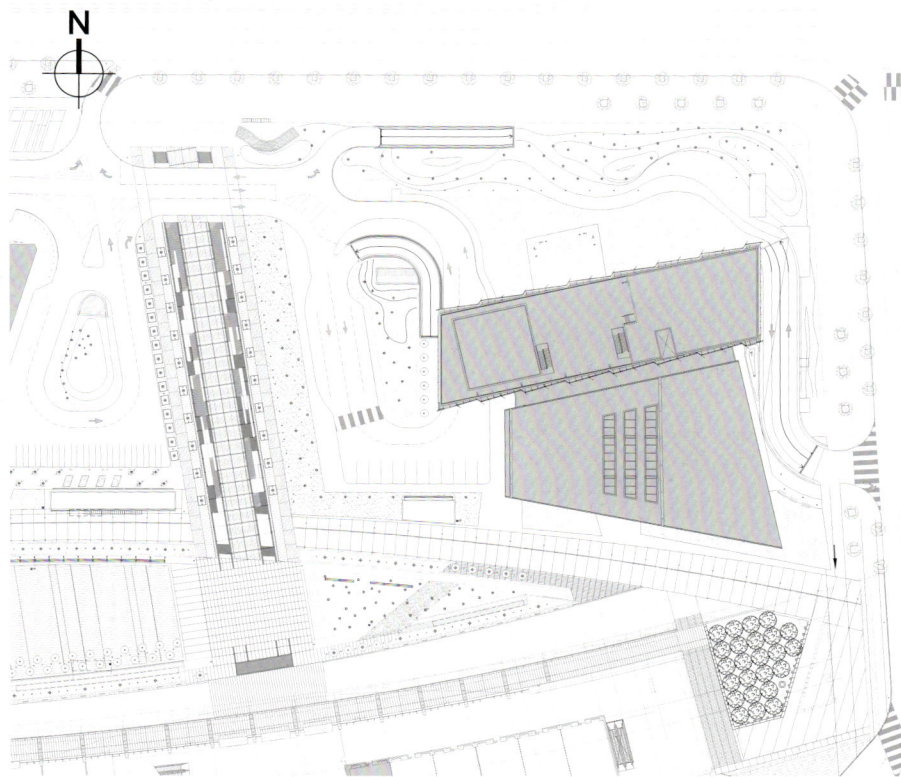

总平面图 Site plan

东北视图（夜景） Northeast view (night)

意大利餐厅 Italian restaurant

中国餐厅 Chinese restaurant

俱乐部休息室 Club lounge

舞厅 Ballroom

仁川喜来登酒店位于松岛国际城的中心，毗邻世界级的设施，比如会展中心、杰克·尼克劳斯高尔夫球场、国际学校和购物中心。在设计一座城市中的商务酒店的想法下，坐落在 400 000 ㎡ 中央公园附近的该酒店为在繁忙的日常生活的你带来一种舒适的放松。

为了实现这一点，建筑外部由现代和高科技幕墙覆盖，考虑到东北亚贸易大厦的天际线和周边的环境，建筑设计为 23 层。幕墙多角度的变化不仅使得这个设计和附近的建筑很协调，并且增加了韵律感，使得它成为环境之中最引人注目的建筑。

Sheraton Incheon Hotel, located on the center of Songdo International City, is in the vicinity of world-class facilities such as a convention center, Jack Nicklaus golf course, international school and shopping mall. It also gives you a comfortable relaxation in the hectic daily life in connection with the 400,000 ㎡ central park under the concept of a business hotel in the city.

To achieve this, the exterior was covered with modern and high-tech curtain walls and the building was designed to have 23 floors considering the skyline of Northeast Asia Trade Tower as well as surrounding contexts. The diverse angle changes of the curtain walls add rhythm to the design to harmonize with buildings in the vicinity, making it the most remarkable architecture among them.

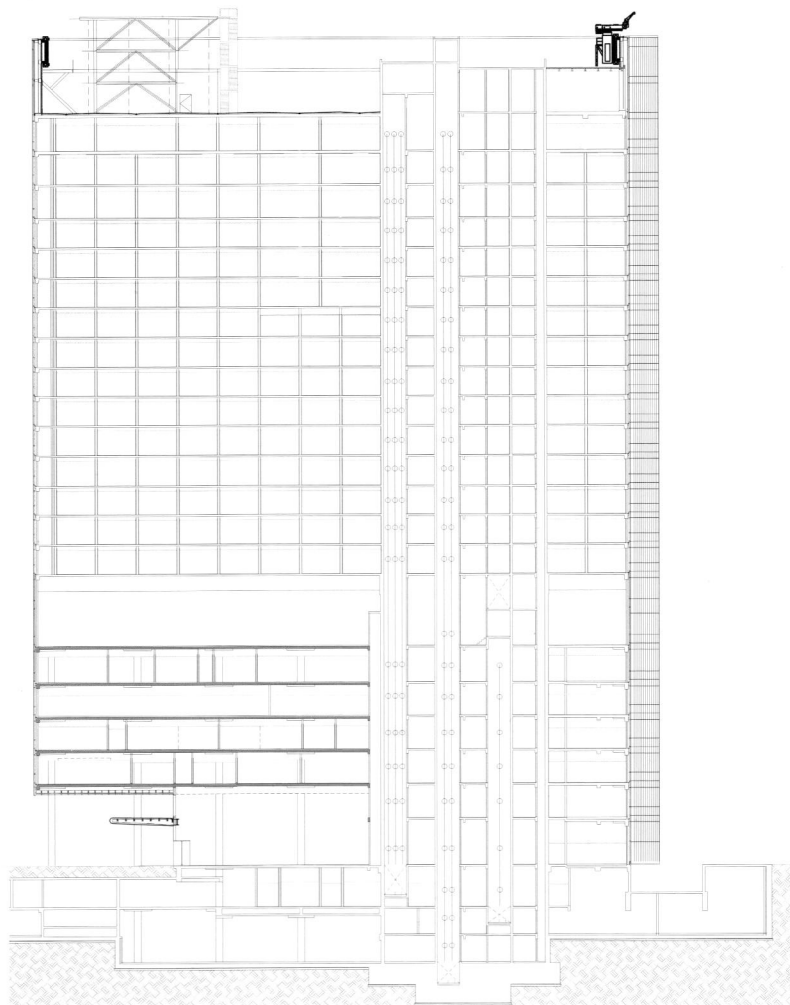

横剖面 Cross section

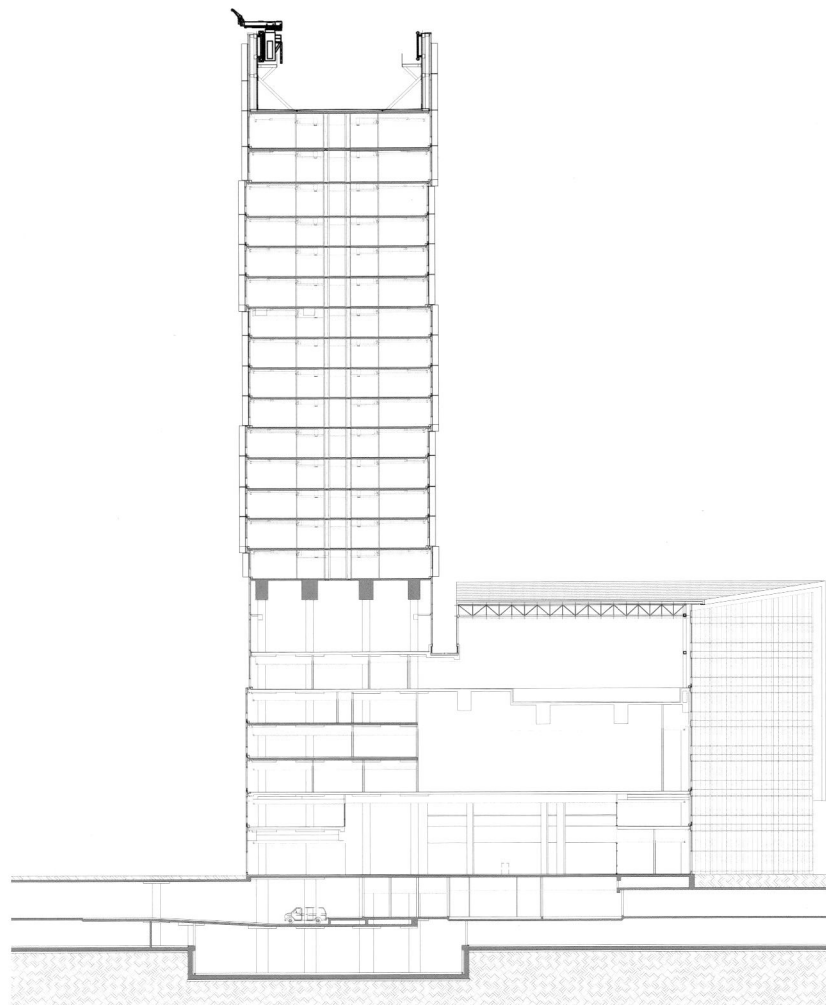

纵剖面 Longitudinal section

国宾房 Ambassador

豪华单人间 Suite Deluxe king

大厅 Lobby

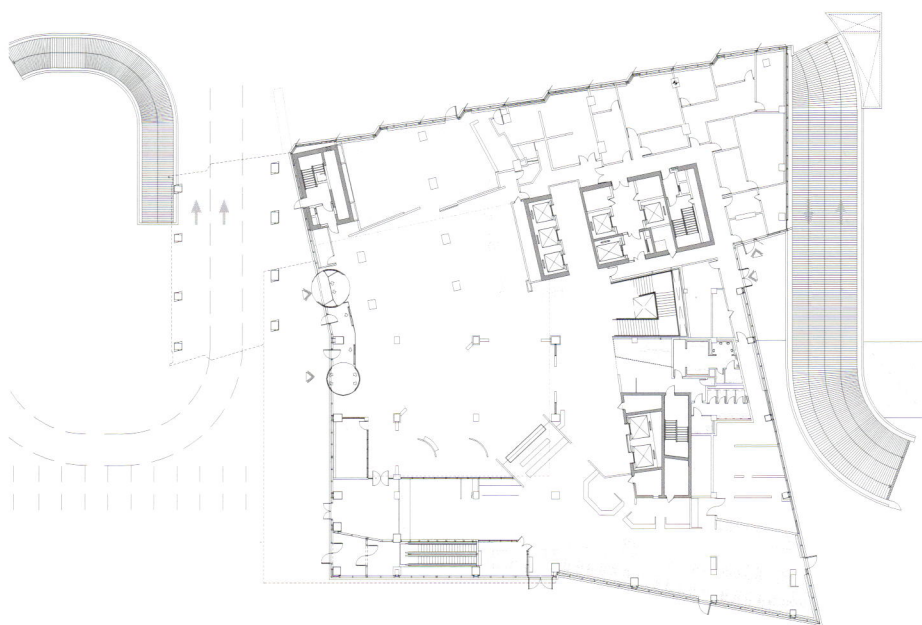

一层平面图 First floor plan

二层平面图 Second floor plan

Rock It Suda
Rock It Suda

© Yum Seung-hoon

设计 moonbalsso / Moon Hoon + Design Network

地点 Kangwon province, Korea

功能 Weekend dwelling

基地面积 2,854m^2

建筑面积 472m^2

总楼面面积 456m^2

楼层 B1, 2FL

结构 Reinforced concrete

饰面材料 T2.3 Urethan painting on steel, Painting on exposed concrete(ext.), Fabric, Tile, Exposed concrete, Compressed floor, Wallpaper, Painting(int.)

摄影师 Declared in picture

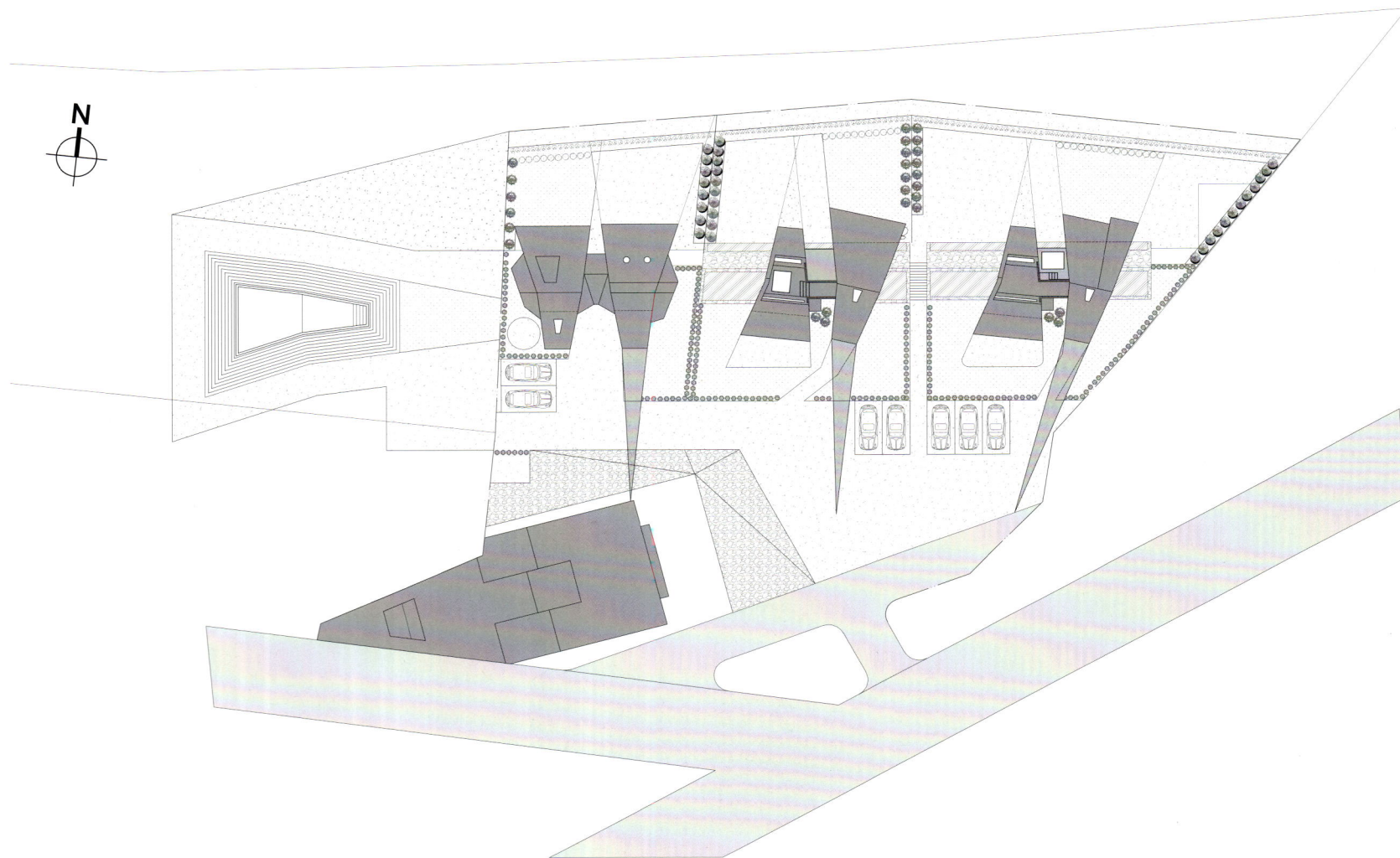

总平面图 Site plan

空间的混合 Mix of space

软性空间 Soft space

空间的扩展与压缩 Extension & Compression of space

色彩与主题 Color & Theme

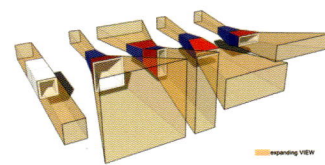

景观 Landscape

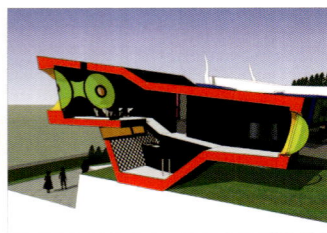

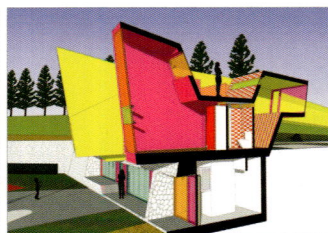

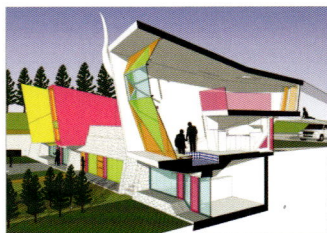

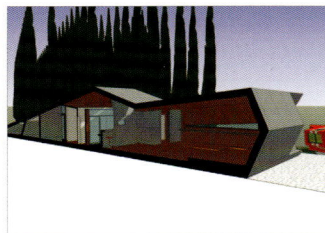

概念图 Concept image

© Yum Seung-hoon
后视图 Rear view

土地和风景

在开始设计之前，当我去项目地点参观的时候，我有很多模糊的预期。我希望这块地能有一个独一无二的灵魂，也许会对这个建筑的基本想法产生一些影响。另一点就是我希望这块地的周边环境是复杂的，那样它们就能对基本的计划产生影响，或者我可以把它们当作设计的力量。总体来说，我个人很喜欢这块不规则形状的土地，有很大的标高差距，或者是受法定限制影响的形状。原因是建筑师对于这样地块的形状的干预最小。重申一下，有意的设计少一点，能够迸发出意料之外的组合。也许我说得太过明显，然而，我认为最终的设计是由最少的意识和最小的意愿来完成的。Ho-chon Ri的土地给我的感觉是圆的、长的，而不是有棱角的。如果我不考虑想象中的分界线，场地前后的风景都很漂亮。我认为，为了在设计中囊括场地前后的风景，我不得不仔细探究。并且，如何囊括这些风景也许是设计的焦点所在。

空间（硬性和软性）

我首先开始的是 33m2 的单元空间的设计（情侣房）。为了使空间有一定的方向性，我选择了长方形的空间，并且决定和场地平行或垂直放置。这样，和这块地平行的空间将会是地下空间，我开始用垂直的空间在空间的扩展和压缩上做一些变化，有四种不同方法来扩展和压缩空间。首先，在水平和垂直的方向上都有一个扩展。然后在水平方向上扩展，垂直方向上压缩，相反地，在水平方向上空间变得更窄小，在垂直方向上空间更宽阔。最后，有一种戏剧性的类型，就是在倾斜的空间上完成对空间的扩展和压缩。在 ROCK IT SUDA 项目中，我倾向于尝试一种新的空间构成方式，"软性"空间。我所说的"软"指的是我计划去创造一个由网组成的空间。比如吊床或是绳网（在孩子们的操场上），让那些网来支撑人们的重量或移动。一部分网可以不必承担人们的重量，可以像一种环境艺术一样在空间扩展的方向上悬垂下来。因为这些装置都可以对风水做出呼应，所以，它们能表达一种"有活力的生活"。

功能

这个设计提供了宽阔的平台和景观，所以用户会有更多的户外活动。客房也有娱乐的功能，并且每一间客房都有自己的主题，因此用户再次光顾时会有新的期待。在类似跑车的客房中，用户能够享受到模拟赛车的感觉。还有以隐形飞机作为主题的空间，内部的空间光线昏暗，并有高差，因此用户能够有一个动态的视野，并且在客房中就能感受到隐形飞机移动的效果。

形状

因为这个空间是从内部开始设计的，所以外部的形状就是内部设计的结果。当它正在寻找一个合适的落脚点的时候，空间的软性部分最终是一个三角形的金字塔形状。自助餐厅所在的建筑是通过皱缩土地登记图上的形状或是旋转不同的轴线（Y轴）来实现的，形状要比人为设计的形状更自然。（最终的版本被改变了，但是仍有早期的精髓。）

ROCK IT SUDA 与牛角

甲方曾去西班牙旅行，似乎斗牛和牛角是他最生动的回忆。当我和一位熟人聊天的时候，"牛角"的想法突然闪现出来。当我跟甲方提到"牛角形"的时候，他带着微笑欣然同意了。一个牛角形的房子就这样诞生了。当这个巨大的牛角被附加于这个建筑的时候，我有一种奇怪的感觉，对我来说它就像是有生命的。

为什么是"TALE"

"TALE"的意思是故事，在英文中，"尾巴（tail）"和"故事"有着相同的发音。Tail 作为动词的时候，有跟随的意思。对这个住宿公寓来说，tale 和 tail 在不同的角度和范围内创造了很多故事。它意味着我们希望很多住宿的客人能够"跟随着尾巴，把酒杯斟满美酒"。我们通过"尾巴"的移动来说明动力。我们谈论作为象征性故事结构的移动的形状，移动到现实的世界中去，比如第二次生命或者是其他数字空间。我谈论现实世界或网络空间的原因是，即使这个设计起来好像是要从现实中逃脱开，或者是无视现实的庄严，但实际上，我想要以一种更加全面、立体的方式，通过改变空间、改变认知去创造一种建筑学的经验。

Land & Scenery

When I visit the project site before I start the design, I have couple of rather vague expectations. I hope the land would have a unique spirit, which may give impact on the basic concept of the architecture. The other is that I hope the surrounding conditions of the land are complicated so that I can use those as good impacts on the basic planning or use those as the energies. In general, I personally like the lands which have irregular shape, which have big level differences or whose shapes are influenced by legal restriction. The reason is that the intervention of architect on the shape can be minimized for such lands. To say again, intentional design could be less, and, unexpected combinations can spring out at this time. Perhaps I am saying too boldly, however, I think the ultimate design is made by minimum awareness and the least amount of will. The land at Ho-chon Ri gave me the feeling that it is round and long, rather than angulated. The sceneries of front and rear were both beautiful, if I would not refer to imaginary boundary line. I thought, "I have to explore the land scenery carefully before containing the sceneries of front and rear in the design". And, I thought that perhaps how to contain the scenery would be the main focus of the design.

Space (Hard & Soft)

I began the design of 33 ㎡ unit space (lovers' house) first. In order to give the directivity of space, I chose rectangular space and decided to place that in parallel to or perpendicular to the land. At this time, the space in parallel with the land will be underground space and I began to give changes in space expansion and compression with vertical space. I had four different ways of space expansion and compression. First, there is the expansion in both directions of horizontal and vertical. Then there is horizontal expansion and vertical compression. Conversely, space can become narrower in horizontal direction and expands in vertical direction. As last, there is a dramatic type where the expansion and compression of space are both in the sloped space. In the space of "ROCK IT SUDA" pension, I intended to try a new way of space form, which was the "soft" space. What I mean by "soft" is that I plan to make a space which is made by "nets", such as hammock or rope-net (in children's playground), and let those nets support the weights and movement of men. For the part of net, which does not have to support the weight of men, that net can be hang down in the direction of space expansion like an environmental art. Since these devices can respond to wind and water, they can express "dynamic life".

Program

The design provides with wide deck and landscaping so that users would have more outdoor activities. The lodging spaces also have entertainment functions and each lodging space has its own theme so that users would come back with new expectation. In the lodging space which looks like a sports car, user can enjoy auto-racing simulation. The lodging space which has stealth aircraft as motif, internal space is dark and there is elevation difference so that user can have dynamic views and movement of stealth aircraft in the lodging space.

Shape

Since the space design began from internal space, external shape became the result of internal design. The space in the soft part ended up having a triangular pyramid shape while it was looking for a proper landing place. The cafeteria building was created by shrinking the shape in the land registration map or by rotating it by different axis (y axis). The shape of the cafeteria building is more natural than intentionally planned shape.(The final version is changed but it still has early energy.)

ROCK IT SUDA & Horn

The client had a trip to Spain and it seems that bullfighting and bull horn were the most vivid memory. The idea of "horn" shape had popped up when I was talking with an acquaintance. When I mentioned about the "horn shape" to the client, he readily agreed to it with big smile. It was like "a horny house was born". When the huge horn was attached to the building, I had quite strange feeling. For me it looks like it still has life.

Why TALE!

"Tale" means a story and "tail" has same pronunciation with "tale". "Tail" as verb means "following". For this lodging pension, "tale" and "tail" makes many stories in various angles and dimensions. It means we wish many lodging guest would come "tail to tail", "bumper to bumper". We say about moving energy through the movement of that tail. We are talking about moving shape as a symbolic story structure, which is moving to virtual world such as second life or other digital space. The reason I am talking about virtual world, or cyberspace, is that, even it may look like the design is trying to escape from reality and ignore the gravity, in fact, I wanted to create an architectural experience in more holistic and three-dimensional way, by varying the space experience, means of recognition or dimension.

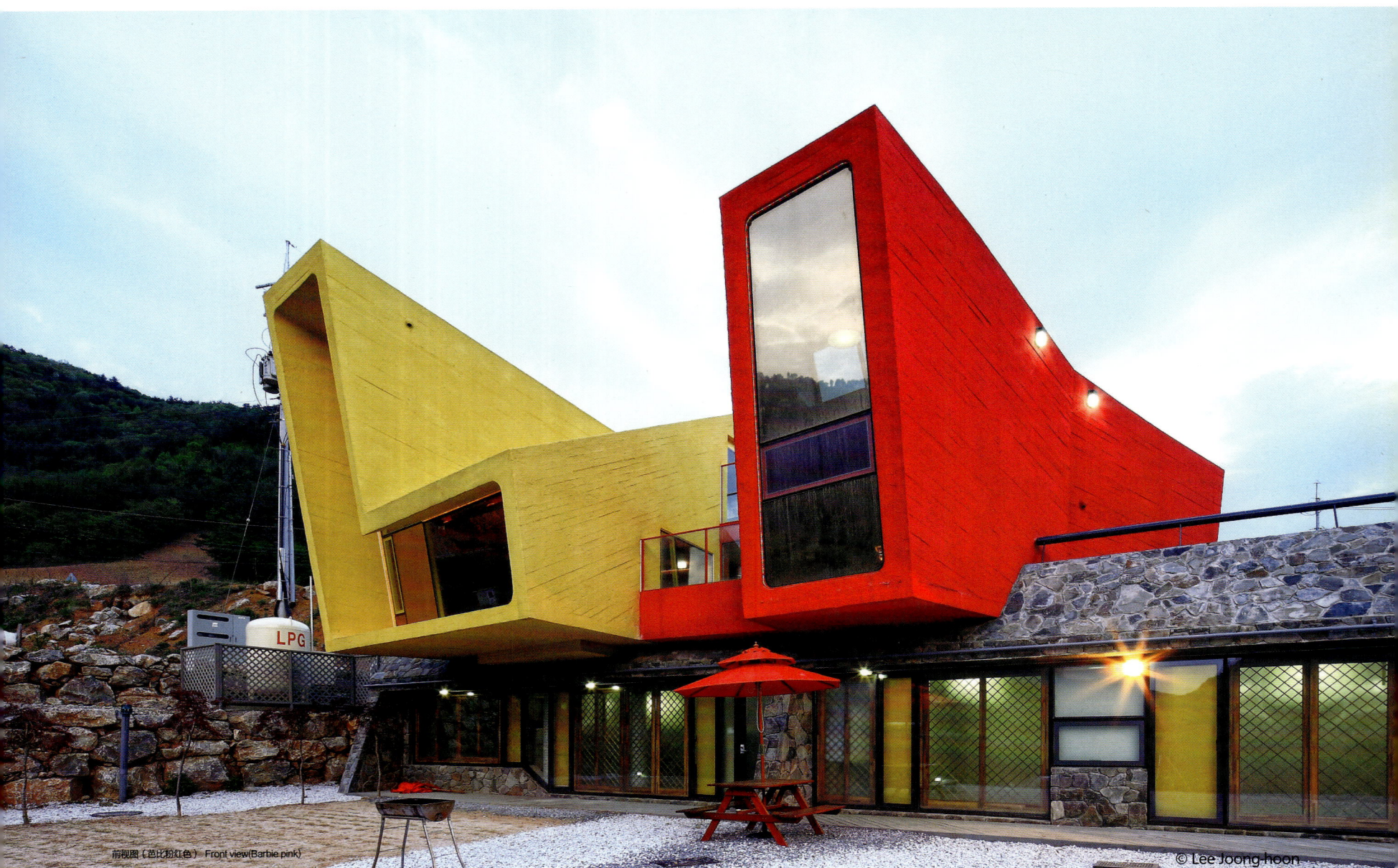

前视图（芭比粉红色）Front view(Barbie pink)

© Lee Joong-hoon

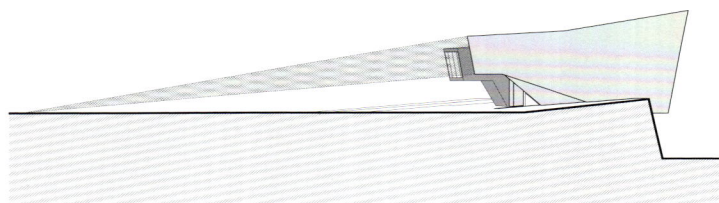

东立面 East elevation

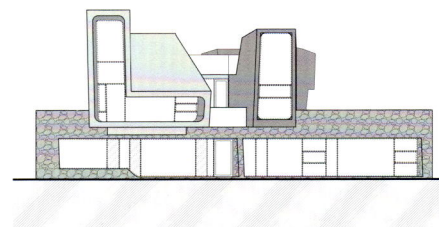

北立面 North elevation

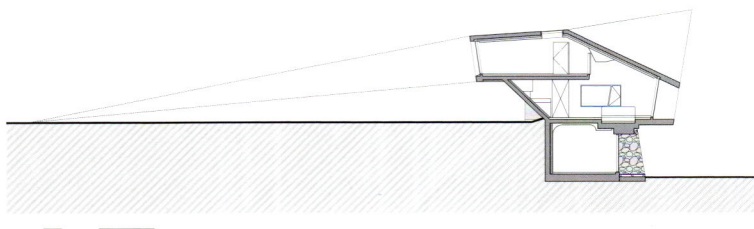

剖面图1 Section I

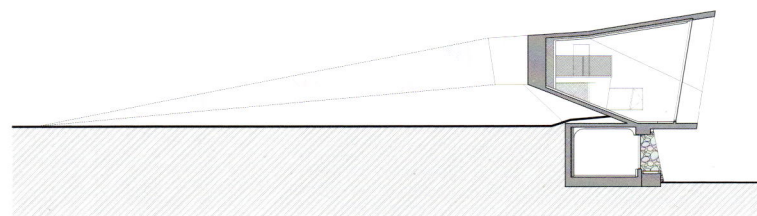

剖面图2 Section II

© Yum Seung-hoon

© Yum Seung-hoon

起居室（芭比的红色） Living room(Barbie pink)

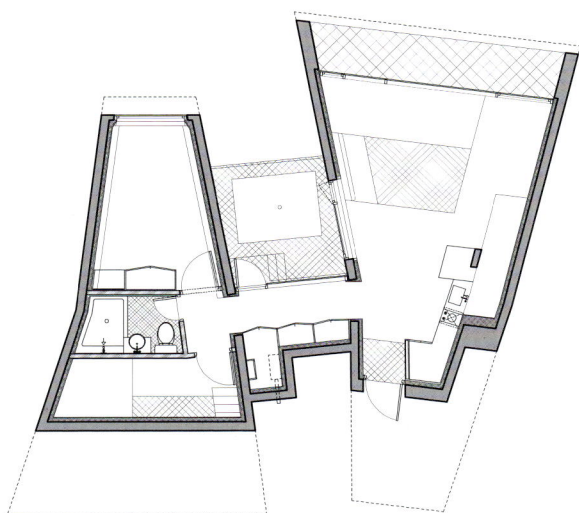

一层平面图 First floor plan

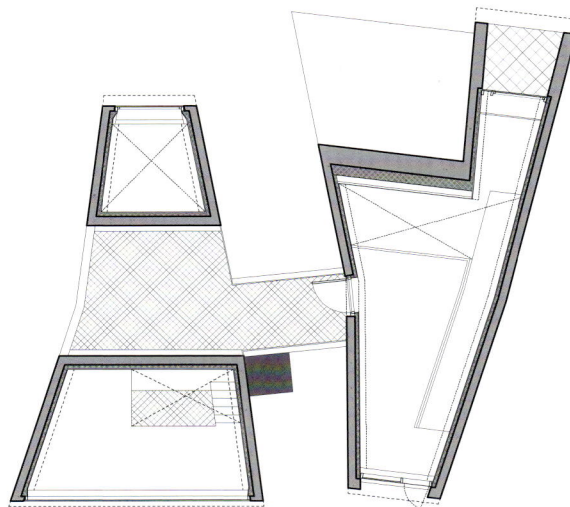

二层平面图 Second floor plan

前视图（西班牙蓝）　Front view(Spanish blue)

© Moonhoon

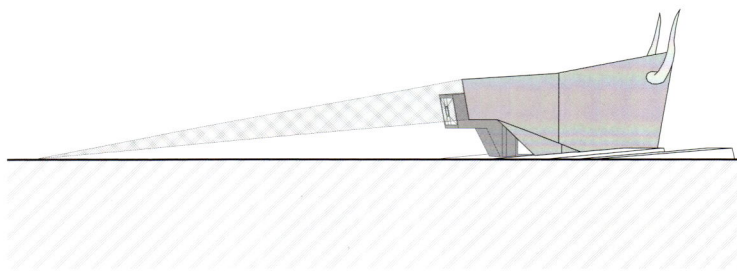

东立面　East elevation

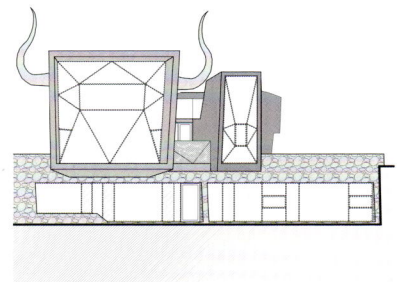

南立面　South elevation

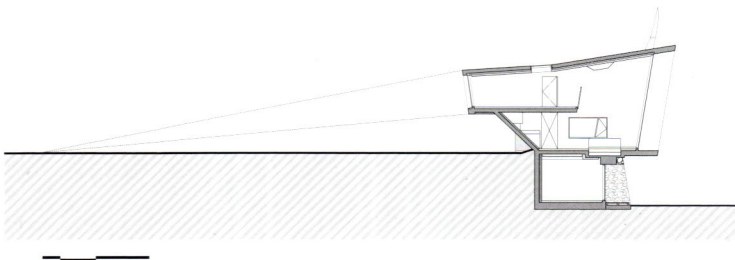

剖面图 1　Section I

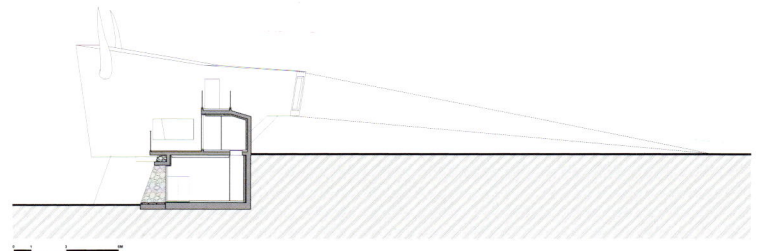

剖面图 2　Section II

© Lee Joong-hoon

卧室 Bed room

© Yum Seung-hoon

起居室（西班牙蓝） Living room(Spanish blue)

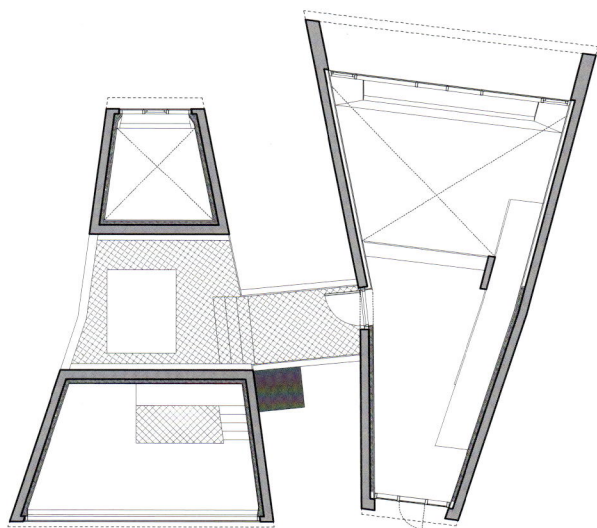

一层平面图 First floor plan

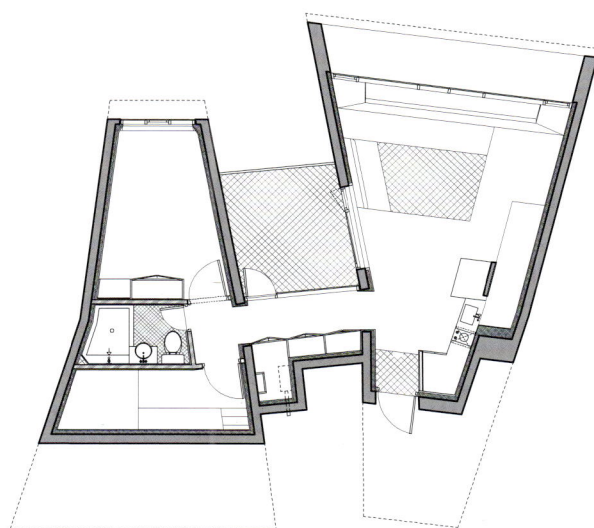

二层平面图 Second floor plan

前视图（隐形飞机和法拉利）Front view(Stealth & Ferrari)

© Lee Joong-hoon

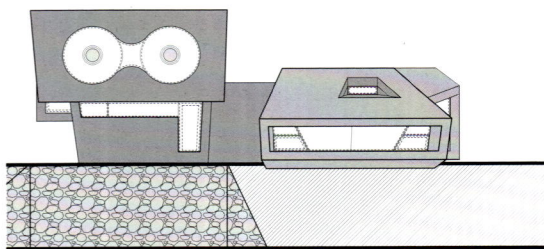

北立面 North elevation

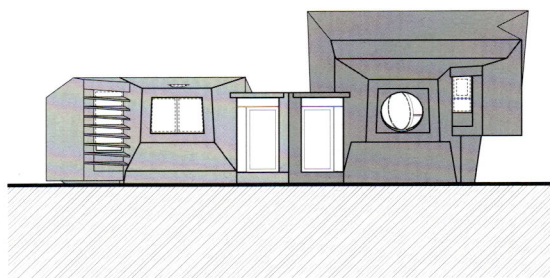

南立面 South elevation

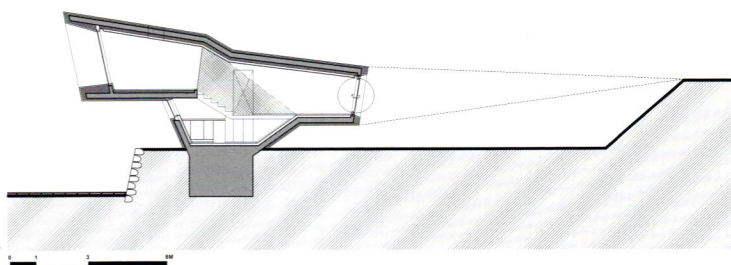

剖面图（隐形飞机）Section (Stealth)

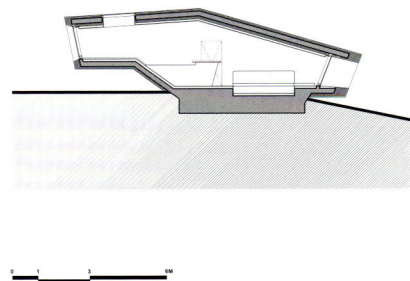

剖面图（法拉利）Section (Ferrari)

© Lee Joong-hoon

© Yum Seung-hoon

© Lee Joong-hoon
隐形飞机）Living room (Stealth)

© Lee Joong-hoon
Living room (Ferrari)

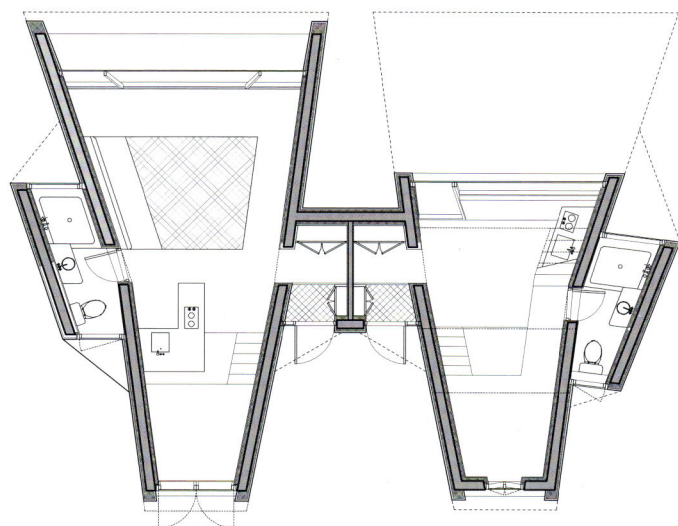
一层平面图 First floor plan

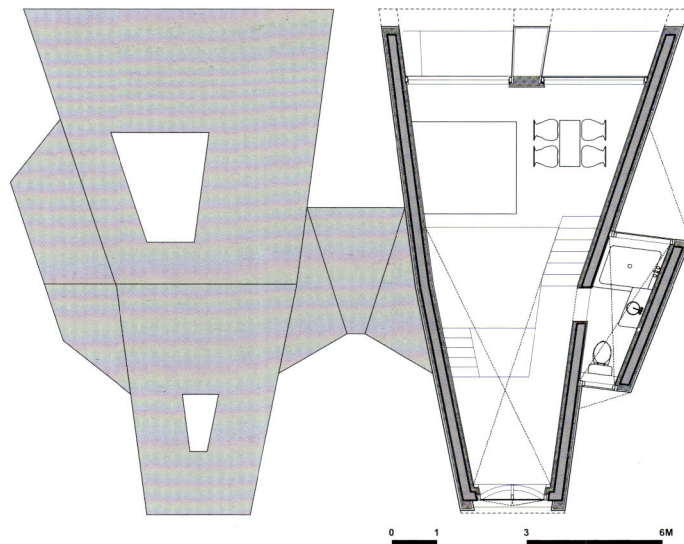
二层平面图 Second floor plan

夜景（Rock it soda） Night view(Rock it soda)

© Lee Joong-haon

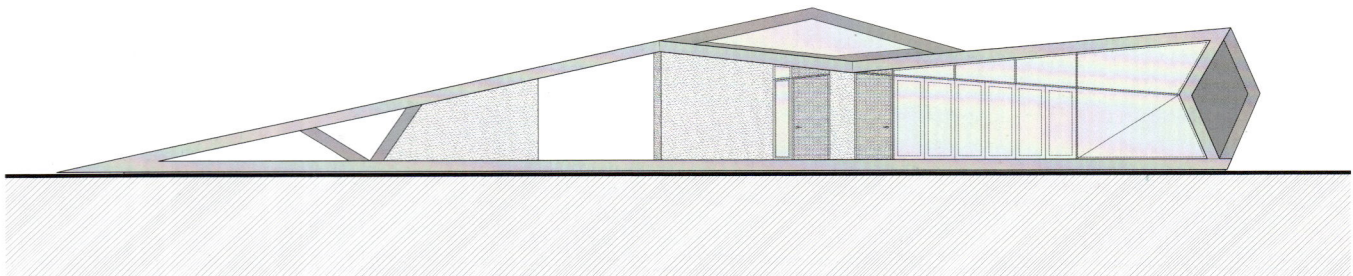

北立面 North elevation

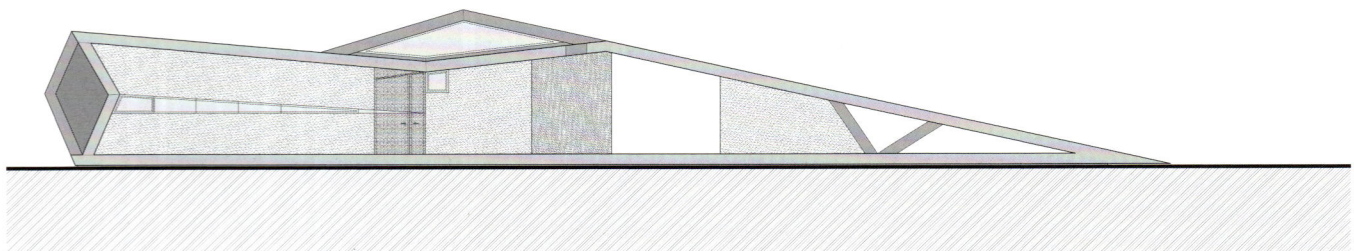

南立面 South elevation

© Lee Joong-hoon

咖啡厅（Rock it suda） Cafe(Rock it suda)

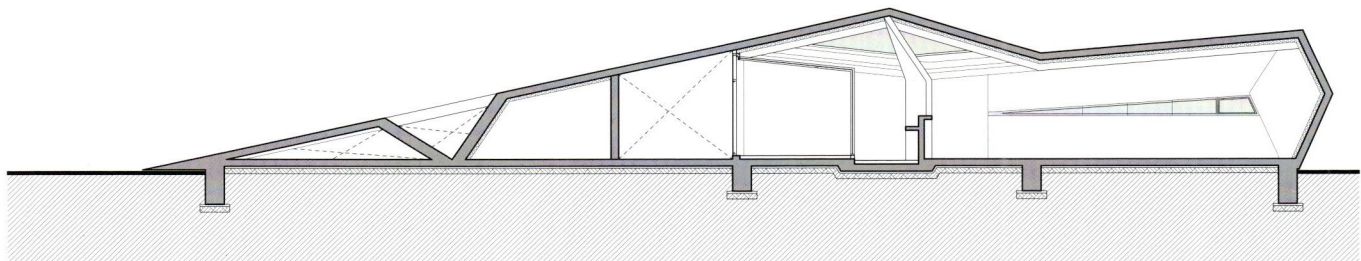

横剖面 Cross section

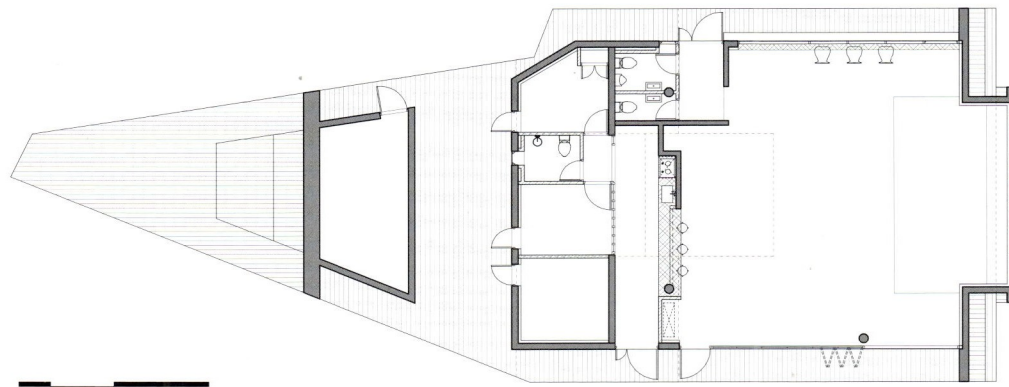

楼层平面 Floor plan

图书在版编目(CIP)数据

度假村与酒店：汉英对照/韩国建筑世界出版社编；
王单单，李硕译.一大连：大连理工大学出版社，
2011.11
 ISBN 978-7-5611-6587-4

 Ⅰ.①度… Ⅱ.①韩… ②王… ③李… Ⅲ.①旅游度
假村－建筑设计－世界－图集②饭店－建筑设计－世界－
图集 Ⅳ.①TU247-64

中国版本图书馆CIP数据核字（2011）第213091号

出版发行：大连理工大学出版社
　　　　　（地址：大连市软件园路80号　　邮编：116023）
印　　刷：精一印刷（深圳）有限公司
幅面尺寸：260mm×300mm
印　　张：32.5
出版时间：2011年11月第1版
印刷时间：2011年11月第1次印刷
出 版 人：金英伟
统　　筹：房　磊
责任编辑：杨　丹
封面设计：卢　炀
责任校对：王单单

书　　号：ISBN 978-7-5611-6587-4
定　　价：388.00元

发　行：0411-84708842
传　真：0411-84701466
E-mail：a_detail@dutp.cn
URL：http://www.dutp.cn